THE FOREST SERVICE
AND THE GREATEST GOOD

THE FOREST SERVICE
AND THE GREATEST GOOD

A Centennial History

James G. Lewis

To Mark — A great friend of The Society and a great supporter of the Hal Rothman Fund.

All the best,
Jamie Lewis

Forest History Society

The Forest History Society is a nonprofit, educational institution dedicated to the advancement of historical understanding of human interaction with the forest environment. The Society was established in 1946. Interpretations and conclusions in FHS publications are those of the authors; the Society takes responsibility for the selection of topics, the competence of the authors, and their freedom of inquiry.

Forest History Society
701 William Vickers Avenue
Durham, North Carolina 27701
(919) 682-9319
www.ForestHistory.org

First edition

Design by Zubigraphics, Inc.

This publication was supported by the Lynn W. Day Endowment for Forest History Publications and supported with additional funds from the USDA Forest Service, New Century of Service, and USDA Forest Service, Office of Communications, in recognition of the centennial anniversary of the USDA Forest Service.

Second printing, with corrections, 2006

Library of Congress Cataloging-in-Publication Data

Lewis, James G. (James Graham), 1965–
  The Forest Service and the greatest good : a centennial history / James G.
Lewis.-- 1st ed.
      p. cm.
  Includes bibliographical references and index.
  ISBN 0-89030-066-6 (hardcover : alk. paper) -- ISBN 0-89030-065-8 (pbk. :
alk. paper)
  1. United States. Forest Service--History. I. Title.
  SD565.L49 2005
  354.5'5'0973--dc22
                              2005021496

To the men and women of the Forest Service and their families

Contents

Foreword *by Char Miller* ix

Preface xiii

Introduction *by Steve Dunsky, Ann Dunsky, and Dave Steinke* 1

Chapter 1: Origins of the Forest Service 6

Chapter 2: Establishing the Forest Service 24

Chapter 3: Powers, Policies, and Fires 56

Chapter 4: The Primacy of Timber and the Promise of Sustained Yield 86

Chapter 5: Recreation, Wilderness, and Wildlife Management in the First Half-Century . . 110

Chapter 6: The Postwar Period and the Rise of the Environmental Movement 136

Chapter 7: New Faces, Changing Values 162

Chapter 8: Traditional Forestry "Hits the Wall" 186

Chapter 9: A New Land Ethic and Ecosystem Management 206

Chapter 10: Reflections on the Greatest Good 232

Appendices

A: Chronological Summary of Events Important to Forest Service History 237

B: Biographical Sketches of the Chiefs of the Forest Service 244

C: Organizational Charts of the Forest Service 250

Further Reading 255

Endnotes 257

Index 277

Foreword

The Forest Service was born in controversy and has yet to escape its birthright. That's a good thing. No public agency, regardless of its age and mission—and 2005 marks the centennial of the Forest Service's commitment to the management of our national forests and grasslands—should be free from public scrutiny, released from public accountability, or able to finesse public critique.

Gifford Pinchot, the agency's founding chief, knew this stricture full well. Arguing in 1907 that the national forests "exist today because the people want them," he declared that the citizenry must assume the primary responsibility for determining their political context and environmental management: "To make them accomplish the most good the people themselves must make clear how they want them run." The public's manifold concerns would not always be in concert, but the social tensions that would result were an essential part of democratic life. Only by acknowledging that controversy was the norm would the Forest Service long endure. That's why Pinchot's most famous maxim was predicated on discord: "and where conflicting interests must be reconciled the question will always be decided from the standpoint of the greatest good of the greatest number in the long run."

The central place of controversy in the history of the Forest Service forms the narrative frame for James G. Lewis's fine synthetic account of the agency's first hundred years. In this, it neatly parallels the organizing theme of a compelling documentary, *The Greatest Good: A Forest Service Centennial Film*, for which it serves as the companion volume.

Like the film, Lewis's book explores these continual conflicts with a critical goal in mind: how to explain the source of the long-standing arguments over the agency's actions? It makes a strong case that this has been the result of the Forest Service's origins. It was, after all, a radical experiment in American political history. When in the late nineteenth century the first federal forest reserves were set aside, their very existence marked a sharp break with past public policy: until then, the General Land Office in the Department of Interior had one primary function— to dispose of the public domain. With the creation of the forest reserves in 1891, their enlargement later that decade, and consolidation and rapid expansion as national forests beginning in the early years of the twentieth century, a new form of public land management was established that was tied to the emergence of a more powerful nation state. In creating an agency that would in time control more than 190 million acres, forester Gifford Pinchot and President Theodore Roosevelt were also extending the clout of the executive branch, shaping the lives and livelihoods of a (largely) western citizenry. Ever since, we have been fighting over the Forest Service's political status and regulatory authority.

That said, it is a wonder that the agency has survived. Lewis chronicles, for instance, the numerous, if ultimately unsuccessful, attempts of politicians, bureaucrats, and interest groups to dislodge or dismantle the nation's first federal bureau framed around conservative land management. In its early years, livestock associations in the West mounted furious attacks upon it. The first Sagebrush Rebellion flared up in Colorado in 1907 over fees charged to use public grasslands (not for nothing did Pinchot dub the battle over grazing the "bloody angle"). Tension between the region and federal land managers has reignited ever since over mining rights, oil and gas exploration, clearcut harvests, and wilderness corridors. Emblematic of this long-running dispute was the spotted owl controversy of the late twentieth century, which pitted loggers against greens, local economies against national environmental legislation, popular belief against scientific findings—and everyone against the Forest Service.

Then there have been the periodic, and highly contentious, attempts to transfer the Forest Service, and the millions of acres under its control, from the Department of Agriculture into the Department of Interior. Albert Fall, who headed Interior under President Warren G. Harding in the early 1920s, was the first to pursue this goal, which his involvement in the Teapot Dome scandal derailed. More famous were Secretary Harold Ickes's strenuous efforts during the Great Depression; those also came to naught, as did discussions during the years when Jimmy Carter and Ronald Reagan held the White House. The attempts tell us a good deal about bureaucratic memory: the nation's public woodlands had originally been under Interior's control but were removed in 1905 when Congress approved their transfer to Agriculture to establish the Forest Service—a transfer that the Interior Department never forgot.

Much less controversial, but no less central to the Forest Service's identity, has been its long-term commitment to landscape restoration. As Lewis makes clear, the lands given to the agency, and those that over time it has purchased, often have been among the nation's most abused—charred, eroded, clearcut, or overgrazed. From its very beginnings, the Forest Service has marshaled its scientists and field staff to repair this devastated terrain, from New Hampshire's White Mountains to the Ozarks of western Arkansas; from the Sierra Nevada to the grasslands of the Great Plains to the Great Smoky Mountains in Tennessee and North Carolina. When in fall 2004 I asked a 92-year-old retired forester, who began work in 1932, about his work during his half-century career in the agency, he laughed: "I planted trees, lots of trees. Maybe a million seedlings." His memory is consistent with the Forest Service's ambitious resolve to mend a broken land.

Its custodial heritage is often forgotten in the wake of its post–World War II focus on timber production. As Lewis reminds us, "getting out the cut" became the agency's mantra in response to the tremendous increase in demand for lumber for the nation's cities, and especially to construct its new suburbs. To boost harvests, it experimented with silvicultural techniques, especially clearcutting. For all the agency's success in delivering a large volume of timber, clearcutting provoked public outcry. The very people whom this wood sheltered in the new communities framed up in the 1950s and 1960s vacationed in many of the national forests then experiencing accelerated cutting; when these suburbanites visited once-beloved woodlands and found them chopped down, they cried foul.

Slow to react to the public's dismay, because it was convinced that its scientific expertise should trump aesthetic concerns, the Forest Service was dumbfounded when it was sued in *Izaak Walton v. Butz* (1973), startled when former allies like the Sierra Club began to criticize its actions, and astonished when the emerging environmental movement denounced the once-reputable agency as the devil in disguise. With public distrust at an all-time high, and with the enactment of a series of landmark legislation, including the Wilderness Act (1964), the National Environmental Policy Act (1969), and the Endangered Species Act (1973), the agency of the late twentieth century found itself in a peculiar position—the regulators were now being regulated.

As he probes these tumultuous events and the Forest Service's reactions to them—some more effective than others—James G. Lewis keeps a close eye on the relationship between these contemporary dilemmas and the past out of which they have grown. Like their predecessors, early-twenty-first-century foresters struggle to locate their generation's greatest good, to find a way to care for the land and serve the people that does justice to both. "There are many great interests on the National Forests, and they sometimes conflict a little," Gifford Pinchot wrote in 1907, but for consensus to emerge it "is often necessary for one man to give way a little here, another a little there." Just how rarely that has occurred during the agency's first century is a measure of how difficult yet essential democratic debate is to the resolution of seemingly intractable environmental problems. Good stewardship requires an open dialogue.

—*Char Miller*

Preface

It is the greatest happiness of the greatest number that is the measure of right and wrong.
— Jeremy Bentham (1776)

All land is to be devoted to its most productive use for the permanent good of the whole people,
and not for the temporary benefit of individuals or companies. . . . [W]here conflicting interests
must be reconciled, the question will always be decided from the standpoint
of the greatest good of the greatest number in the long run.
— Gifford Pinchot (1905)

What Pinchot adds is "for the longest time." That's what foresters do. They think out across time
and that time shapes what the first part of the phrase is involved with.
Whose greatest good is it now? Whose greatest good will it be later?
And might their vision of what constitutes great be different from the beginning?
— Char Miller (2003)

The practice of conservation must spring from a conviction of what is ethically and esthetically right
as well as what is economically expedient. A thing is right only when it tends to preserve
the integrity, stability, and beauty of the community, and the community
includes the soil, waters, fauna, and flora, as well as the people.
—Aldo Leopold (1947)

I think if you look at the greatest good for the greatest number in the long term and then you turn and
you look at [Leopold's] land ethic, that looks to be an evolutionary process, not a dramatic breakthrough.
— Jack Ward Thomas (2004)

When the transfer of the federal forest reserves from the Department of the Interior to the Department of Agriculture occurred in February 1905, Gifford Pinchot was ready. As chief of the U.S. Division of Forestry in the Agriculture Department, he had spent seven years politicking, persuading, courting, and cajoling to make it happen. The wording and structure of the statement above about how to reconcile conflicting interests is classic Pinchot, as is the story of how it was issued. It appeared in a letter from Secretary of Agriculture James Wilson addressed to Pinchot dated the same day as the transfer. Pinchot had ghostwritten the mission statement for Wilson to sign, and arranged that it would be issued agency-wide the day Pinchot's agency took charge of the federal forests.

The statement was short, direct, to the point—and borrowed. Pinchot had adapted Jeremy Bentham's statement of utilitarian philosophy to his own needs. Pinchot made a career of adapting and expanding the ideas of others and pushing them in new directions. For all of his creativity, Pinchot was not known as an original thinker. He was, however, a great synthesizer of ideas, an inspiring leader of men, and a brilliant politician. Pinchot engineered one of the great legal land grabs in history. At the urging of his friend, President Theodore Roosevelt, Congress moved 63 million acres of public land from one department to another and placed Pinchot in charge of it.

Arguably, the more impressive part of this story is that Pinchot did it for largely altruistic reasons. The United States was in another high cycle of graft and corruption at the time of the transfer, and the Department of the Interior, where the forest reserves had been, had a well-documented history of land fraud and poor management. Instead of going to Interior where the land was, Pinchot wanted a fresh start for federal land management and arranged for the land to come to him.

It is a rare instance of the mountain coming to Mohammed—and indeed Pinchot possessed the zeal and purpose of a prophet. He chose his missionaries carefully and assembled the men and machinery to deliver the message. He made use of government franking privileges to get the word out about forestry and conservation and the purpose of the Forest Service. His forest rangers were well versed in the message because the Forest Service manual bore the above statement about the greatest good for the greatest number for the longest time on its very first pages. It has been the guiding statement of the agency ever since.

A majority of the early forest rangers had learned that statement from Pinchot himself, or at the Yale Forest School, which he had funded to train men for the conservation mission. One of those men was Aldo Leopold, who, like Pinchot, adapted existing ideas and broke new ground with his words and actions. Leopold and his family labored for many years to rehabilitate a piece of exhausted land in Wisconsin's sand counties area. The work gave Leopold a new perspective, and it gave him time to reflect on his experiences. All of this came together in his collection of essays, *A Sand County Almanac,* where he shared his ideas about a different way to look at the land, and to manage it. Two decades would pass before these ideas were widely disseminated in the mainstream, and nearly another two passed before they affected how the Forest Service managed public land. When the agency adopted "Ecosystem Management" in the 1990s and altered its approach to land management, it did so with the goal of restoring the land, much as Leopold had begun more than fifty years earlier in the sand counties and Pinchot had begun fifty years before that on a cut-over, burned forest in North Carolina: each in his own way, each for his own reasons, each for the greatest good.

The story of the Forest Service and the greatest good is not only the story of Gifford Pinchot and Aldo Leopold. It is the story of those two men, and of the thousands of other men and women who have worked for the agency in its several incarnations for more than one hundred years. And it is a story of myriad conflicts, and of what arose from those conflicts. *The Forest Service and the Greatest Good: A Centennial History* is not a comprehensive history of the agency. It would be impossible to tell the entire history of the agency in these pages, or in ten times as many pages. Each individual who worked or volunteered for the agency has his or her own version of the agency's history to tell. It is to the men and women—past, present, and future— of the Forest Service to whom this book is dedicated.

Instead, this is *a* history of the Forest Service. In 1899, Gifford Pinchot published *A Primer of Forestry* in order to introduce the public to forestry and the mission of the U.S. Division of Forestry. It is hoped that this book serves a similar purpose—that it introduces the public to the Forest Service, to the issues it handles, to its successes and its mistakes, to its triumphs and its tribulations. The agency is not, nor has it ever been, monolithic in word and deed. Over the years, the individuals who have worked there have disagreed with one another and sometimes with the public they serve. Through the exchange of ideas, however, the Forest Service has learned and even prospered, and, as a consequence, the land has benefited from this exchange.

Something similar can also be said of the experience in writing this book. Through numerous conversations and interviews, and from delving into the literature on the Forest Service, I have learned and prospered as a historian. While I cannot begin to list all those who granted me interviews, I must thank Karla Hawley, assistant to Chief Dale Bosworth, for arranging an interview with the chief, and thank Chief Bosworth for speaking with me so candidly. That opportunity convinced me to carry the history on up to the centennial year.

Regarding the literature, anyone writing about the Forest Service must start with the work done by Harold K. "Pete" Steen. First published in 1976, his *U.S. Forest Service: A History* remains the best place to start for the administrative history of the agency. His countless books and collections of edited papers on various aspects covering Forest Service and public land management history are invaluable as well.

I also learned and prospered from Jerry Williams, the venerable national-level Forest Service historian who was a great boon to the film and this book. Jerry freely shared his valuable time, his many articles, artifacts, and photos, and his unrivaled encyclopedic knowledge of and insights into the agency. Jerry was also one of the manuscript reviewers, and the book is better for it. Char Miller and Steve Anderson, the former a superlative mentor and the latter an understanding and patient boss who kept my writing on point, did valuable service as reviewers. John Fedkiw of the Forest Service also gave generously of his time and institutional knowledge as a reviewer, providing comments that were helpful with minor revisions in this reprint. Dave Steinke and Steve Dunsky, the co-directors of the film who thankfully wanted a companion book, also reviewed the manuscript and made valuable suggestions. They and their staff members Mario Chocooj and Judy Dersch and Ruth Williams of Animal Ocean provided many of the photos and the artwork from the film for the book. Adele Logan Alexander, Amanda Burbank, Frank Carroll, Gary C. Chancey, Kim Ernstrom, Wes Farris, Tom Iraci, Scott Jones, Lori Messenger, Margarita Phillips, Becky Philpott, Susan Stewart, Michael Williams, and Wayne Williams, to name a few, also provided images and data on tight deadlines. One additional word about Steve Dunsky. By telephone and email, Steve provided numerous vital ideas and supplied several contacts, sometimes while in the middle of a critical time in the filmmaking process. He and Ann, his wife and the film's editor, kindly housed and fed two wayward souls one weekend during the writing process. Of greater importance, they have become good friends.

Thanks go to Elizabeth Hull and Michele Justice and especially the tenacious Cheryl Oakes, of the Forest History Society, for their research assistance, and to Cheryl for her helpful feedback. Mary Braun provided significant behind-the-scenes aid at the end. Jeannie Conner whipped the manuscript into great shape, patiently wading through numerous drafts and providing excellent feedback. Sally Atwater did an outstanding job of polishing it at the end, and Kathy Hart at Zubigraphics of putting text and images together in an engaging way. Dianne, my wife, cheerfully and with a keen eye for detail looked over and discussed sections of the book and helped with the index. Despite all the eyes and marking pencils that went over the manuscript, I take full responsibility for what is contained within.

Lastly, thanks to my parents and family and friends, who have helped me along the way. I would especially like to thank Dianne, who has been extraordinarily patient and supportive, giving up precious vacation time and weekends to travel with me on book- and film-related trips. I've learned a great deal about the craft of writing over the years from her and have prospered from sharing this experience with her.

—*Jamie Lewis*
June 2005

Introduction

Forest Service history is a messy business.

We discovered this when we made the documentary *The Greatest Good*; James Lewis did as he began to write this book. Our problem: how to fit the diverse dimensions of the Forest Service into a single coherent story.

Reviewing the literature, we found this to be a persistent problem in Forest Service historiography. Authors who choose a chronological structure jump from one resource area to another and then repeat the pattern in the next era. If they choose a topical structure and discuss grazing, water, timber, and fire in turn, they fragment the agency's story until it becomes a collection of disconnected parts.

Lewis melds these approaches as he seeks to create a unified narrative. A fine example is his look at women in the Forest Service. In Chapter 7, "New Faces, Changing Values," he looks at the role of women over several decades. He places the issue in the larger context of how Forest Service culture evolved from the early days right up to the 2001 appointment of Associate Chief Sally Collins, the second-highest-ranking person in the organization. It's not about milestones in women's history but about the struggles they faced working in a male-dominated organization *and* as Forest Service employees trying to adapt to changing demands on the agency itself.

The discussion of women in the agency points to another problem we had in making the film. With so many topics and issues to cover, our solution in editing the film was to omit important information. We favored comprehension over comprehensiveness. Our objective was to hold the audience's attention for two hours. Since the content of a feature-length film is roughly equivalent to that of a novella (about a hundred pages), we knew that we would not cover many critical topics in Forest Service history. We saw the need for a companion text from the beginning and decided to work with the Forest History Society to make that happen. We are delighted that this volume fills so many gaps left by the film.

We are particularly excited that this book touches on topics that we encountered in researching and shooting the film but have not been detailed in other books about the Forest Service. One of the more fascinating aspects of recent Forest Service history was the agency's involvement, at several levels, in the Vietnam War. It is only within the past few years, as files have been declassified, that this story could be told. And so it is presented here as part of the change the Forest Service was undergoing in the late 1960s.

Lewis has also incorporated quotations from the hundreds of hours of interviews that we shot for the film, and from oral interviews with figures who have long since passed on. This is wonderful material that simply did not fit into our program but goes a long way toward illuminating aspects of the agency's history by the individuals who lived it.

When we began *The Greatest Good* project, the subtitle for our documentary was "The History of the Forest Service." We soon renamed it "A History of the Forest Service" and finally "A Forest Service Centennial Film." This progression illustrates our increasing awareness of the complexity and depth of the information, and our decreasing comfort with the notion that we were telling the whole story. We now believe that it is impossible to tell the whole story—and perhaps there is no "whole story."

Like the film, *The Forest Service and the Greatest Good* does not try to tell the whole story, but rather seeks to introduce readers to the many accomplishments and trials of the Forest Service over the past century in hopes that they will begin to appreciate the complexity of the agency's history and mission and seek out more information on their own. The goal of both the film and the book is a balanced look at the agency that can encourage all those who care about our national forests and grasslands to engage in a dialogue about just what is the greatest good.

The film and companion book offer two more perspectives, in different media, on this corner of conservation history. We hope you will find them to be significant contributions, but we also encourage you to explore the range of histories that consider various aspects of the Forest Service. Collectively, they tell a part of this big sprawling story.

—Steve Dunsky, Ann Dunsky, and David Steinke

National Forest System

National Forests
National Grasslands

Regions of the Forest Service

Chapter One

ORIGINS OF
THE FOREST SERVICE

Sorting logs at Glens Falls, New York, ca. 1890. (P007026—Courtesy of The Adirondack Museum)

EVERYTHING, IT SEEMED, WAS MADE OF WOOD. IF IT WAS NOT MADE OF WOOD, THEN IT CAME FROM THE WOODS, OR THE WOODS HAD BEEN CUT TO MAKE WAY FOR IT. THIS WAS—AND IN MANY WAYS STILL IS—A NATION BUILT OF WOOD. FROM COLONIAL SETTLEMENTS UNTIL THE EARLY TWENTIETH CENTURY, AMERICANS MOVED ALONG WOOD ROADS AND SIDEWALKS, LIVED IN WOOD HOMES, SAT ON WOOD FURNITURE, AND BURNED WOOD FOR COOKING AND HEATING. BERRIES, MUSHROOMS, MAPLE SUGAR, AND OTHER PRODUCTS CAME FROM THE WOODS, TOO.

Settlement and development altered the landscape. Settlers opened up forests to make way for planting wheat, corn, and other crops. By the 1850s, the dense northeastern forests lauded and celebrated by Transcendentalist writers and Hudson River landscape painters had become open fields or cleared mountains. One historian estimates that by 1850, at least 100 million acres of land had been "improved," or cleared, usually for agriculture. It took only sixty more years to clear an additional 190 million acres, nearly double what Americans removed during the previous two hundred years.[1] Fire, grazing farm animals, and poor agricultural practices contributed to soil erosion. The loss of fertile soil made farming more difficult, leading many to abandon their land and move on to clearing more forests, thus perpetuating the cycle. Displaced topsoil made its way into waterways, where the silt complicated navigation and compromised water supplies, as well as affecting fish and wildlife populations.

The nineteenth century was also the era of cut-and-run lumbering. A typical logging operation cleared nearly all timber with no regard for regeneration.[2] There seemed to be less water afterward, and many people noted how shallow nearby streams and rivers became. Loggers took only merchantable wood to market and left the slash on the ground, creating a fire hazard. Because land was taxed, timberland owners did not hold onto the land long enough to harvest a second crop. Instead, they either sold it to farmers or defaulted on the taxes, which returned the land to the public domain, and moved on to the next forest. With a seemingly endless supply of timber, little thought was given to future needs.

Loggers followed settlers westward from Maine and New England in the early 1800s, to the Great Lakes region by the Civil War, to the South by the 1880s, and then to the Pacific Northwest. Lumbermen exploited loopholes in laws like the Homestead Act of 1862 and the Southern Homestead Act of 1866, both of which were intended to encourage settlement, to acquire more timberland on the cheap. Railroads, which opened up these lands to settlement, needed large amounts of timber to make railroad ties and fuel steam engines. Without wood, iron itself

The new technologies of log hauling and transport transformed and devastated forests. Steam-powered skidders like the one in the background, developed in the 1890s, used chains to drag logs along the ground and could load an average of 125,000 board feet a day. The process left almost no trees for natural regeneration. (Forest History Society)

could not be made. The Iron Horse, with all its clanging metal and mechanized power, had a wooden heart.[3]

Until the Civil War, though, few Americans had given much thought to how the demand for wood might affect the land or the economic future of the country.[4] The lumber industry had long experienced boom-and-bust cycles, and the post-Civil War period proved no different. It was an unstable time for the lumber industry: while the price of land and timber remained constant, the cost of manufactured lumber fluctuated. Fear of a timber famine and anxiety over diminished water supplies generated concern for the future of American civilization. Civic leaders and leading intellectuals argued that uncontrolled wood consumption would ultimately destroy the economy and then society itself, and bring the return of savagery.[5]

European states had begun developing forestry as a science in the late 1700s in response to wood shortages and watershed problems. Instead of discarding slash as Americans did, European foresters used every part of the tree, down to scraps the size of pencils.[6] Germans took the lead in developing scientific forestry and utilitarian management that ensured steady supplies of both timber and water and profits for the landowner. Only after comparing conditions between Europe and the United States did Americans make the connection between deforestation and watershed protection.

George Perkins Marsh

The shift from passive concern to intervention on behalf of American forests began with George Perkins Marsh's book, *Man and Nature: Or, Physical Geography as Modified by Human Action*. Published in 1864, Marsh's study of European ecology and human impact on the land profoundly and deeply influenced thinking on the subject in the latter half of the century. His book—a history of the destruction of Europe's forests with theories on the interconnectedness of land, water flow, and forest cover—went through numerous reprints and was widely read throughout the English-speaking world in the nineteenth century. It remains an important work and is still in print, and Marsh's findings still stir debate.[7] Nineteenth-century forest preservationists and forestry advocates such as Charles Sargent and Gifford Pinchot closely read this work and based their arguments for government intervention on it.[8]

Marsh focused on ecology, or the relationship between humans and their environment, and drew his conclusions initially from his observations in the United States and abroad. As a Vermont farmer and, later, as ambassador to Italy and Turkey, he saw firsthand the effects of deforestation by humans and grazing animals, the damage inflicted on mountains and waterways by destructive and irresponsible forest practices, and the effects of soil erosion on surrounding areas. He observed how soil washed down from the overcut and overgrazed hillsides into rivers and streams, clogging and flooding Vermont's rivers and streams and triggering landslides and flooding in Europe's valleys.[9]

George Perkins Marsh, author of *Man and Nature*, provided the intellectual foundation for the conservation movement. Gifford Pinchot, who was deeply influenced by Marsh's ideas, called the book "epoch-making." (Library of Congress)

Marsh viewed humans as a new geological agent in the world because of their power to change the earth drastically. Given that much power, he argued, citizens and their governments must assume moral responsibility toward the land.[10] Marsh concluded that the survival of civilizations depended upon proper management of watersheds.[11] He advocated stewardship of the land on a global scale in a time when the U.S. government was doling out large land grants without regulating activity on the public domain.[12] No advocate of preservation, Marsh hoped to see large-scale planning for the utilitarian management of forests for the sake of soil and water quality. He recommended scientifically managing the forests, creating tree farms, and harvesting only mature timber. He also called for managing the land through draining, damming, and irrigation. Forty years after he published these ideas, Congress established the Forest Service within the Department of Agriculture to implement these practices.[13]

Marsh also discussed whether planting trees would alter the climate and increase rain. Even though he came to no conclusion, congressmen familiar with his arguments chose selectively from them to support their causes and ignored the call for rational management.[14] They passed a spate of federal laws aimed at promoting the semiarid West as a new Garden of Eden. Government scientists with the Smithsonian Institution, Department of Agriculture, and other agencies promoted the theory that planting trees increased rainfall, and Arbor Day was created in 1872 to publicize this belief.

Indiscriminate logging and watershed damage motivated conservationists to create the National Forest System. On what would become the Wasatch National Forest in Utah a year later (above), clearcutting and overgrazing as seen in 1905 endangered this watershed. Mining interests near Leadville, Colorado, left nothing but stumps on what is now the Rio Grande National Forest (right). (USDA Forest Service)

Congress, caught up in the excitement of opening more public land for settlement, embraced the idea of afforestation (converting open land into a forest by planting) to help the family farmer. Based on the Homestead Act of 1862, which gave 160 acres of public land to any citizen who was the head of a household and over age twenty-one, the Timber Culture Act of 1873 gave 160 acres to any qualified citizen who would cultivate trees on 40 acres and keep them healthy for ten years. Through the Desert Land Act that same year, the federal government sold land for $1.25 an acre to any settler who agreed to irrigate it. However, few farmers had the means to take advantage of these offers and succeed. Instead, the Timber Culture Act and the Desert Land Act were exploited by cattlemen, who used them to acquire large tracts for grazing.[15]

State legislatures and private organizations also took action to preserve forests on public lands. Concern over the diminishing availability of high-grade timber prompted several states to adopt tree-growing bounties and tax exemptions. Local organized efforts to protect woodlands began in earnest in 1871 in the wake of the worst wildfire in the nation's history. At Peshtigo, Wisconsin, more than fifteen hundred people lost their lives and nearly 1.3 million acres of forests burned.[16] Two years later, the American Association for the Advancement of Science (AAAS) began agitating for forest preservation by calling for minimal development to protect watersheds. The efforts of AAAS led to the creation of the American Forestry Association (AFA) in 1875. Forest preservation, not the scientific management Marsh had observed in Europe, became the initial goal of the two groups.[17] Only a minority of members looked to the European forestry systems, which employed trained foresters to manage protected forests, as their model.

Few Americans knew that for more than a century, European foresters had been harvesting timber in a manner that produced new trees for future cutting and produced a profit for the landowner, all while protecting watersheds. Most Americans concerned about the woodlands assumed that anyone who wielded an ax—even foresters who practiced partial cutting methods that promoted regeneration—could not be trusted. Support for total preservation led the New York State legislature to stop the sale of state woodlands in the 1870s, to create the publicly protected Adirondack Forest Preserve in 1885 and, finally, to add the "forever wild" amendment to its state constitution in 1894. No development is permitted on the state-owned land within the Adirondack Preserve to this day. The idea of preservation, though, angered western settlers and their congressional representatives when advocates presented preservation bills on Capitol Hill in the 1880s and 1890s. California established the first state forest board in 1885 but limited its activity to education and research. Preservation meant no usage, and most westerners wanted no part of that.

The Division of Forestry

Much like the congressional representatives who selectively borrowed from Marsh's book to justify their legislation, advocates for the Adirondack Preserve had drawn from government statistics and reports published in the 1870s and 1880s to support their position. Those favoring scientific management drew on the same reports to support their position. Most of the reports came from the desk of one man, Franklin Hough.

Hough, a central figure in the establishment of AFA and AAAS, was a physician by training who had a strong interest in forestry. His publication of local histories of lumbering activities in three counties in his home state of New York led to work for the U.S. Bureau of the Census in 1870. His analysis of Census reports revealed that lumber production was falling off in some

Franklin Hough (left), the first chief of the U.S. Division of Forestry, advocated a strong federal policy for the protection and management of federal lands. His successor, Nathaniel Egleston (right), was equally sincere and conscientious in promoting forest protection but was a weak administrator. He was relieved rather than upset when the more qualified Bernhard Fernow replaced him. (USDA Forest Service)

areas and increasing in others, indicating to him that lumbermen were exhausting timber supplies. He began gathering scientific data and statistical information in the 1870s and pushing the federal government to protect forests in the public domain. As an intellectual pursuit, he began collecting data on forests in both the United States and Europe around the same time.[18]

At the 1873 AAAS meeting, Hough presented a paper, "On the Duty of Government in the Preservation of Forests," which drew heavily on Marsh's ideas in *Man and Nature*. Hough appealed to both preservationist and scientific forestry sentiments, and the paper inspired the organization to take action.[19] As the group's chair, Hough wrote to President Ulysses S. Grant about the cultivation of timber and the need for preserving forests. He summarized the problems facing American forests and called for a government-funded forestry investigation to assess the quantity of lumber, as well as its rate of consumption and waste throughout the country; to examine how forests affect temperature, rainfall, and other climatic conditions; and to provide a full statement about European forestry methods and schooling.[20] Finally, Hough called upon Congress and state assemblies to pass legislation protecting the remaining forests in the public domain and acknowledging the economic benefits of preservation.[21]

Representative Mark Dunnell of Minnesota, a member of the House Public Lands Committee, wanted Hough to carry out the work. After failing on several occasions to get a forestry bill passed, Dunnell appended the substance of the bill to a general appropriations bill. The amendment allotted $2,000 to the Department of Agriculture for a man of "approved attainments" to compile a report on the condition of American forests. As a result of this last-minute maneuver, Congress placed a forestry agency in Agriculture, which had no woodlands under its control.[22]

Appointed as the first federal forest agent in August 1876, Hough was humbled yet filled with the "oppressive sense of the magnitude of newly acquired burdens." He had little need to worry—he had been collecting information for five years from Europe and in the United States and had a running start on the project. Funding, though, was a different matter. Hough knew it would require the "strictest economy" to prepare the report.[23]

Congress kept Hough on a short fiscal leash, just as it would do with many of his successors.[24] One of the major problems throughout the Forest Service's history has been low salaries. Esprit de corps and a love of the great outdoors might feed a ranger's soul, but a decent income puts food on the table. Living conditions for forest rangers fared little better because of limited budgets for government-financed housing. Those posted to the central office in Washington in the early twentieth century fondly remembered the unlimited cold milk, baked apples, and gingerbread at meetings of the Society of American Foresters held at the Pinchot home—often the best meal they might eat for a week. Twenty years later, things had not improved much. A survey of 210 ranger stations in 1920 showed only forty-six with running water and three with bathtubs.[25]

Hough published three forestry volumes in 1878, 1880, and 1882. Together, they provided the most comprehensive information available on American forest conditions and their history, the timber industry and foreign trade, and recommendations for action.[26] His conclusions echoed Marsh's: the destruction of American forests was widespread and had to stop; government ownership of the land seemed the only practical way to halt wasteful lumbering; and implementation of forestry practices on a national scale that included regeneration of trees for perpetual harvesting (later called sustained yield) offered the best hope for averting economic and environmental catastrophe. One historian credits Hough with pioneering two critical breakthroughs in American forestry: the concept of sustained yield and the system of national forests that emerged by 1900.[27]

Impressed by Hough's reports, Congress established the Division of Forestry within the Department of Agriculture and appointed him its first chief in 1881. He worked tirelessly to promote his views and compile information for the cause of forestry. Nonetheless, a personal rivalry with his new boss, Commissioner George B. Loring, ended in Hough's demotion in 1883, which left him as an assistant to his successor, Nathaniel Egleston. Two years later, Hough finally quit; he died shortly thereafter, in June 1885. Egleston served for three ineffective years and left the division in worse shape than when he took over.[28] Hough's demotion would not be the last time a chief of forestry would be punished for clashing with an administrative official.

Charles Sprague Sargent

Shortly after his appointment as forestry agent in 1876, Hough received support from his friend, Secretary of the Interior Carl Schurz, whose knowledge of the condition of American forests proved a boon to the nascent forestry movement. Appointed in 1877, the German-born

In addition to his long career at the Arnold Arboretum and his impressive reports on forests and trees, Charles Sprague Sargent also founded and edited the highly influential magazine *Garden and Forest* from 1887 to 1897. The publication combined practical and scientific information for gardeners with discussions of forestry problems. (Courtesy of President and Fellows of Harvard College, Archives of the Arnold Arboretum)

Schurz had seen the benefits of scientific forestry in his own country and agreed with Hough on the need for federal governmental action. Consequently, the secretary placed original documentation about American forests in the 1880 Census report, including a catalog of trees and a map of the forests.[29] In Charles Sprague Sargent's *Report on the Forests of North America (exclusive of Mexico)*, the ninth volume of the 1880 Census, issued in 1884, Schurz got what he wanted. The report also established Sargent's reputation as the leading authority on American trees and forests in the country. He later enhanced that reputation with *The Silva of North America*, a catalogue of trees found in the United States and Canada published in fourteen volumes between 1890 and 1902, standard references still in use today.

A Boston Brahmin and a graduate of Harvard College in 1862, Charles Sprague Sargent learned the basics of horticulture when he took over management of the family estate while in his twenties.[30] Appointed as the first director of Harvard University's Arnold Arboretum in 1872, he had questionable qualifications for the position, but he was a quick study and was mentored by Asa Gray, the preeminent American botanist of the day.[31] Sargent developed the Arnold Arboretum into a world-famous center for the study of trees while becoming a leading botanist in his own right. He served as director for fifty-four years.

For most nineteenth-century Americans, the word forester conjured up images of the folk hero Robin Hood and his Merry Men roaming the royal forest, righting wrongs and helping the poor. The image of Robin Hood as a forester remained so strong in the American mind that some early Yale forestry students formed the Robin Hood Society of Foresters in 1905, a private social club whose members dressed in period costume to emulate their fictional hero.[32] The American scientific community, meanwhile, considered trees and forests the realm of botanists and forest biologists (or plant ecologists), not foresters—a natural assumption in a country with no scientifically managed forests and only one trained forester, Bernhard Fernow, who could not even find professional employment. Almost all tree studies and forest investigations in the United States had been botanical studies—cataloging or counting tree species, or making timber estimates. The 1870 Census had included a section on forests, a first for the Census, but its author, William H. Brewer of Yale College's Sheffield Scientific School, relied on existing literature and second-hand oral reports.[33] In contrast to the slow but steady growth of data about individual species, knowledge of the extent and number of forests remained vague, limited, and generalized.[34] In the United States, there were no studies of the forests as part of a larger ecosystem, no comprehensive first-hand investigations, and no studies of forest management. Few people in the country, it seemed, could see the forests for the trees. Sargent's appointment in 1879 to oversee the Census project would change that.[35]

Bernhard Fernow, the third chief of the U.S. Division of Forestry, was intimately involved in drafting several forestland laws, including the Forest Reserve Act and the Organic Act. As chief forester, he laid the foundation on which the present organization of the Forest Service was built. (USDA Forest Service)

In midsummer 1883, Sargent finished his thorough report, which drew heavily from Marsh's work.[36] In addition to its catalogue of North American trees, the report included an examination of Maine to see what had happened after the passing of its lumber heydays. Sargent found scientific forest management to be practicable "when the importance of the forest to the community is paramount." He continued: "The forests of Maine, once considered practically exhausted [in the 1820s and 1830s], still yield largely and continuously, and the public sentiment which has made possible their protection is the one hopeful symptom in the whole country that a change of feeling in regard to forest property is gradually taking place."[37] Once the community found scientific management essential to its material prosperity, Sargent argued, it would then take action. Here was the paradigm for bringing conservation to the rest of the country: an informed public, or at the very least, one facing disruption of the source of its economic livelihood, would want to change its habits—an idea suggested seven years before as part of Hough's first report.[38] Sargent's work served as the standard reference on the subject for the next twenty years.

Fernow Takes Charge

Work on the Census report kept Sargent from attending the meeting of the American Forestry Congress in April 1882. The meeting marked the beginning of Bernhard Fernow's forty years as a crucial presence in both American and Canadian forestry.[39] That Fernow has received so little recognition for his role in organizing and fostering the forestry profession in America speaks

Split-rail fences, just one of many uses for wood by settlers, required enormous quantities of lumber. A rail fence to enclose a square forty-acre field needed about eight thousand fence rails. By 1850, there were some 3.2 million miles of wooden fence in the United States.
(Forest History Society)

more to the efforts of his professional rivals to expunge him from the historical record than it does to his actual contributions.[40]

Born in Prussia in 1851, Bernhard Eduard Fernow trained in both law and forestry to manage substantial family agriculture and forest holdings.[41] Instead of fulfilling his family's wishes, he traveled to the United States in 1876, ostensibly to attend the American Forestry Association meeting held in Philadelphia in conjunction with the centennial celebration. While there, the young forester married his American-born fiancée and became a U.S. citizen.[42] He found work running an iron furnace and consulting for the iron works on its forested property in Pennsylvania, which afforded him a chance to study northeastern forest conditions. Though he did not bring the fifteen thousand acres under active forest management, Fernow based his plans for forestry in the United States on what he learned in the Pennsylvania woods.[43]

Fernow first gained attention at the American Forestry Congress meeting in 1882 with his paper on the forest policy of Germany. (The American Forestry Congress and its rival, the American Forestry Association, merged that same year and eventually became American Forests.) As the only trained forester in the country until 1890, and as executive secretary of the American Forestry Association from 1883 to 1895, and chief of the Department of Agriculture's Division of Forestry from 1886 to 1898, he played a significant role in the forestry movement.[44] Fernow corresponded with a wide variety of people, encouraged interest in forestry through AFA's periodic meetings around the country, and helped draft state and federal legislation to preserve and manage public land.

Fernow's appointment as the third chief of the Division of Forestry in 1886 brought a shift in the division's fortunes and purpose.[45] Shortly after he took office, Congress gave full statutory recognition to the division, which seemingly ensured its future. Moreover, his appointment placed a forester in charge of the division for the first time and marked the beginning of forestry as a government science.[46]

The new chief faced a difficult job. Fernow had no land to manage, let alone on which to conduct research. Congress had established the division in 1881 largely to advise farmers and agriculturalists wanting to know more about trees and shrubbery. Its focus was trees, not forests. Even after granting the division full statutory recognition on June 30, 1886, Congress continued to deny the division adequate funding, thereby hobbling many of Fernow's efforts.[47] Fernow remained the only trained forester on the payroll during his twelve years as chief. Nathaniel Egleston, Fernow's ineffective predecessor, served as his assistant.[48]

Fernow struggled to reorient the division from office work toward field research. He tapped experts around the country to contribute to scientific forestry literature while carrying out his own laboratory experiments.[49] Because the Division of Forestry had no control of forests, Fernow tried to arrange forestry work through other divisions or on private land—unsuccessfully. In 1890, he prepared the country's first large-scale management plan for a private forest owned by the Adirondack League Club, but he was not involved in the work on the ground.[50] At the insistence of his political bosses, he spent much of his time answering inquiries from casual gardeners and horticulturalists and sending out seed packets. Furthermore, Congress ordered him to conduct frivolous rainmaking experiments in the name of agricultural progress. The experiments served only to embarrass him and the division, as well as to take him away from his more practical experiments.[51]

Nevertheless, Fernow worked diligently for the greater good. During his twelve years as chief, one historian noted, "he firmly implanted in American forestry the idea that a supply of wood was fundamental to civilization."[52] He also argued that unless investors learned that forest management measures could return a profit, forestry would never be accepted.[53] That effort began almost immediately after his becoming chief. In his first annual division report, he called for the withdrawal of public lands from sale, both to protect watersheds and to use them to demonstrate the economic benefits of good lumbering practices. His plan for a forest service to administer the forests, and for reorganizing public land into districts of twenty to thirty reserves, with each reserve comprising ten thousand to twenty thousand acres, was adopted by his successor.[54]

Fernow's reports played a pivotal role in publicizing the cause of scientific forestry.[55] As chief, he published more than six thousand pages of reports, bulletins, circulars, and other materials—more than his two predecessors combined. He worked with Congress to draft legislation because he knew that without congressional support there could be no federal forestry program. While laying the groundwork for federal forestry, he expanded the size and budget of the Division of Forestry, and he and his researchers contributed to substantial new understanding of American trees and wood properties. By the time he left office in 1898 to teach forestry at Cornell University, Congress had enacted legislation creating a national forest system and authorizing its management.

For Fernow, though, introducing forestry to the United States was a bittersweet achievement. He grew weary of the political nature of his job and the politics of the forestry movement. He became so pessimistic that he left Washington convinced that the federal government had prematurely taken on the task of managing forests. Believing that forestry must be implemented on a local level to demonstrate its potential, he envisioned Cornell's school forest in the Adirondack Mountains as an ideal laboratory. Instead, the clearcutting and artificial regeneration efforts he undertook came under such harsh criticism that the state withheld funding and forced the school to shut down in 1903, only five years after it had opened. Disheartened, he moved to Canada in 1907 to start that country's first forestry school. His reputation in the U.S. in tatters, his endeavors largely unappreciated, he continued contributing to forestry through his teaching and a constant stream of publications until his retirement in 1919.

The National Forest Commission

In 1890, the American Association for the Advancement of Science asked that President Benjamin Harrison seek measures to protect the public domain "for the purpose of insuring the perpetuity of the forest cover on the western mountain ranges, preserving thereby the dependent favorable hydrologic conditions."[56] After several failed attempts to pass a separate law, Section 24, a rider supported by Secretary of Interior John Noble, was quietly attached to the General Land Law Revision Act of 1891 just before a congressional recess. The Forest Reserve Act, as it became known, granted the president authority "from time to time" to set aside as "public reservations" any public lands forested or with undergrowth. It did not provide, however, for the management of the reserves, nor did it allocate money to protect the land.[57]

Less than a month after signing the act, President Harrison created the first forest reserve. The Yellowstone Park Timberland Reserve included land around the southern edge of the first national park. By the time he left office in 1893, Harrison had created fifteen reserves containing more than thirteen million acres. Initially, a water-starved West welcomed the Forest Reserve Act. Southern California residents made several adamant requests for General Land Office investigations that led to the creation of forest reserves to protect watersheds in that region.[58] Harrison's successor, Grover Cleveland, added five million acres but then stopped. He saw no reason to continue if the government did not also provide the means for protecting the forest reserves from unlawful entry. Congress failed to pass forest management legislation over the next few years.

Wolcott Gibbs, head of the National Academy of Sciences, decided in 1894 to circumvent Congress and proposed a national forest commission to investigate the public lands and formulate a management policy. Fernow opposed further investigation but decided a unified front was the best way to institute forest management.[59] He and the young American forester Gifford Pinchot drafted a letter for Secretary of Interior Hoke Smith to sign, requesting that the academy, as scientific adviser to the government, launch an investigation and submit a report outlining a "rational forest policy" for the United States. The report would address the practicality of fire prevention in the forest reserves and determine the best way to administer them.[60]

To serve on the National Forest Commission, Gibbs in 1895 selected Academy members William Brewer of Yale, director of the 1870 forest census; General Henry L. Abbot, formerly chief engineer of the U.S. Army; Arnold Hague of the Geological Survey; and Alexander Agassiz of Harvard, a renowned naturalist and oceanographer. He named Charles Sargent chairman. Sargent did not ask Fernow to join them. He had long despised Fernow and the activities of the Division of Forestry, and believed Fernow lacked ability.[61] He even expressed concern that Fernow was working to sabotage the appointment of the commission.[62]

Instead of Fernow, Sargent requested as the commission's secretary Gifford Pinchot, whose career in forestry Sargent had fostered. Pinchot, the first American-born forester, had trained in Europe and had spent the previous five years as a private forester working up and down the eastern seaboard for wealthy patrons. At thirty, he was less than half the age of most of the commission members, but he was eager to examine the forest reserves and get them under scientific management. The commission would travel to many of the reserves and other public timberlands, and it was an opportunity of a lifetime for the young forester.[63]

Pinchot considered Sargent's attitude about the prospects of introducing practical management and about the forestry movement in general as, quite simply, "glum."[64] The very definition of a young man in a hurry, Pinchot would have no part of that way of thinking. He and his friend Henry S. Graves, who was studying to be a forester, set out for the West six weeks before the rest of the commission and began studying forest conditions in Montana. Naturalist John Muir joined them for part of the trip. Pinchot grumbled that Sargent was unwilling to venture far from where the railroads could take them, and preferred hotels to camping.

President Cleveland had requested that the commission produce a plan for federal forest management and provide a list of additional forest reserves to include in his December message to Congress by November 1, 1896. The commission submitted no plan but instead recommended the creation of thirteen new forest reserves. On Washington's Birthday in 1897, Cleveland created thirteen reserves, comprising more than twenty-one million acres, without consulting with representatives of the states affected and without indicating whether resources in these new reserves would be available for use.

When word of the "Washington's Birthday" reserves got out, westerners exploded. Several western congressmen drafted bills to overturn the reserves, one of which Cleveland pocket-vetoed. Denying the current timber needs of the settlers and miners in the area was an affront to the westerners, who saw the move as an attack on individual and property rights.[65] Still other westerners were angry that the reserves still had no protection. In response to the outcry, members of Congress sympathetic to forest preservation joined with scientific authorities in

The Boone and Crockett Club

Organized by Theodore Roosevelt and his friends and relatives in January 1888, the Boone and Crockett Club was one of many sportsmen's clubs that supported game and land preservation efforts in the late 1800s. Members were to be avid outdoorsmen, and not armchair naturalists. Its upper class, influential members included forester Gifford Pinchot and Arnold Hague of the U.S. Geological Survey. Other sportsmen who subscribed to the club's principles of forest protection included Presidents Chester Arthur, Benjamin Harrison, and Grover Cleveland, Secretaries of Interior John Noble and Carl Schurz, and foresters Bernhard Fernow and Carl Schenck. Cofounder George Bird Grinnell used his influential *Forest and Stream* magazine to promote the club's agenda, which included passage of the 1894 Yellowstone Park Protection Act and the Forest Reserve Act of 1891.

working toward a politically and scientifically viable bill to resolve the problem.[66] While they researched and debated, western resentment about being shut out from the public lands grew.

Meanwhile, the National Forest Commission was preparing its final report, and the debates over its content exposed the tensions among members of the commission. The commission's failure to submit a forest management plan at the end of 1896 had especially bothered Pinchot. He blamed Sargent for western opposition and the failure to make clear that the reserves were for the rational use of the forests. He also viewed other aspects of the report as injurious to scientific forestry. With the support of Hague from the Geological Survey, Pinchot threatened to submit a minority report. Despite its flaws, however, the report would focus national attention on the misuse of the public timberlands.[67] A divided commission would weaken the cause. Pinchot eventually bowed to political reality, and though he later regretted it, in May 1897, after Cleveland had left office, the National Forest Commission filed its full report with Pinchot's signature.[68]

The report emphasized that preserving the forests not only protected the water supply, it made "systematic and intelligent forest reproduction" possible. That, in turn, meant that a sustainable timber supply could meet future demand for lumber and other wood products.[69] Accordingly, the federal government needed to act as soon as possible. The committee ended the report with proposals on administering the public domain. Wise forest management, the report concluded, called "for technical knowledge which must be based on a liberal scientific education," and for men above reproach. The U.S. Cavalry was one answer. The Army already patrolled the national parks and could do the same in the forest reserves. Forest officers could be trained at the U.S. Military Academy at West Point.[70] During the few years while the new forest service underwent training, the commission said, the reserves should remain "locked up" and unavailable for development.

Bernhard Fernow's worst fears had been realized. When the president announced the new reserves but failed to provide a management plan, it left the fate of the reserves hanging.[71] Fernow and others persuaded Senator Richard Pettigrew of South Dakota, a powerful member of the Senate Public Lands Committee, to submit an amendment to the new appropriations bill. The Pettigrew Amendment, better known as the Forest Management Act or Organic Act, passed in June 1897. The Organic Act carefully laid out the purpose of the forest reserves and how they would be managed.[72]

The Organic Act stated in part, "No public forest reservation shall be established, except to improve and protect the forest within the reservation, or for the purpose of securing favorable conditions of water flows, and to furnish a continuous supply of timber for the use and necessities of citizens of the United States"; but it was not intended to include lands "more valuable for the mineral therein, or for agricultural purposes, than for forest purposes." In other words, forest reservations protected watersheds and provided the basis for sustained-yield management of forest products and services. In a move designed to minimize criticism from westerners, the

reserves excluded agricultural and mining land. Moreover, mining was specifically allowed in the forest reserves under existing public mining laws (such as the General Mining Law of 1872) and national forest guidelines.[73]

The amendment also directed the secretary of Interior to make rules and regulations for the protection of the reserves "against destruction by fire and depredations" and permitted the secretary to sell "dead, matured, or large growth of trees" in part to promote "the younger growth" in the forests. This was the focus of forest management as understood at that time: it was all about timber—raising, removing, and replanting a crop of trees. The law further instructed that timber to be sold must "be marked and designated, and shall be cut and removed" under supervision. The requirement to mark trees for removal pleased Fernow and Pinchot, who hoped that it would prevent clearcutting on public lands.[74]

Pinchot's participation in the debate over the forest reserves was far from over. He accepted an offer from the secretary of Interior to serve as a "confidential forest agent" working with the Geological Survey's Division of Geography and Forestry to examine the reserves and then prepare a practical plan for the management of the forests—the same work the National Forest Commission had been assembled to complete. A furious Sargent denounced him as a traitor, then walked away from politics and the forestry movement for good.[75]

The federal government initially deployed the U.S. Cavalry to protect Yosemite and other national parks while Congress debated who should manage national parks and forests. Charles Sargent, John Muir, and others favored training Army officers as foresters, a proposal opposed by civilian foresters Gifford Pinchot and Bernhard Fernow. (National Park Service)

Pinchot's plan differed from that of the commission's in calling for a civilian forest service. The rigorous qualifications and expectations Pinchot outlined as necessary for employment were identical to those required of his men when he became chief of the Division of Forestry and its successor, the Forest Service. His basic organization plan—dividing the country into seven districts, each under the supervision of a district forester who administered the land according the natural and market conditions of the region—was very similar to what he later implemented, and nearly identical to what Bernhard Fernow had proposed twelve years earlier.[76]

The fight with the National Forest Commission and the tussle to pass the Organic Act had soured Fernow—no less than Sargent—on the government and politicians. Weary from more than a dozen years of fighting for the cause of forestry, disenchanted with the movement, and feeling unappreciated by the federal government, Fernow accepted a position as director of the New York State College of Forestry in April 1898. Thus began America's first four-year degree program in forestry at Cornell University.

In his *Report upon the Forestry Investigations, 1877–1898*, a government bulletin issued eight months after he resigned from civil service, Fernow dismantled the old arguments against forestry management voiced by forest preservationists. Moreover, the report made a strong case for the Agriculture Department's involvement in forestry. Noting that the time had come not only to "more vigorously pursue technical investigations" and to undertake systematic management as "all other civilized nations apply to their forest property," he declared the work the proper domain of Agriculture's Division of Forestry.[77] It was now up to his successor, Gifford Pinchot, to make that happen, but Pinchot would have to wrest the job away from the Interior Department.

Even in his valedictory moment, however, Fernow could not escape another slight. Secretary of Agriculture James Wilson, who had prevailed upon Pinchot to succeed Fernow, minimized Fernow's accomplishments while trumpeting the work of his newest chief. In his letter of transmittal submitted with Fernow's report to the president, Wilson desired "to call special attention to the fact" that since Pinchot had taken over, "the work of the Division has been directed in distinctly different channels." He emphasized the work discussed in Pinchot's annual report for 1898, including the inception of forestry on private lands, the need to reduce the loss from forest fires, and Pinchot's appeal for an increased budget. Wilson concluded his letter, "These plans meet with my full approval."[78]

With Pinchot came other progressive conservationists who shared his faith in the federal government's ability to implement policies based on research, data, and measurements of American environmental conditions rather than on personal observations. Men like W J McGee, whom Pinchot called the "scientific brains" behind conservation; Henry Graves, a leader in forestry education and Pinchot's right-hand man; and Frederick Newell, chief of the Bureau of Reclamation and architect of the irrigation work, worked closely with Pinchot and President Theodore Roosevelt. This new generation of government scientists would study forests as well as individual trees; they

Secretary of Agriculture
James Wilson (left) helped
persuade Gifford Pinchot
(right) to take charge of the
Division of Forestry in 1898
and then let Pinchot run
the agency as he saw fit.
(USDA Forest Service)

would create policies for watersheds, not just political districts; and they would seek to control those who profited from the land, not forfeit control for the profit of a few.

Some things did not change, however, with Gifford Pinchot's appointment to head the Division of Forestry. His office still had no forests to manage. The forest reserves were in the Interior Department, and management tasks were divided between the General Land Office and the Division of Geography and Forestry. To secure the future of federal forestry, Pinchot decided his first task must be either to go to the forests or to bring the forests to him. Never one to shrink from a challenge, he chose to do both at the same time.

Chapter Two

ESTABLISHING
THE FOREST SERVICE

A forest ranger on the Cabinet (now Lolo) National Forest, 1907. (USDA Forest Service – Forest History Society)

G IFFORD PINCHOT'S APPOINTMENT AS CHIEF OF THE DEPARTMENT OF AGRICULTURE'S DIVISION OF FORESTRY HERALDED THE IMMINENT ARRIVAL OF PROFESSIONAL FORESTRY TO THE UNITED STATES. HIS FAMILY'S WEALTH AND STATUS FURNISHED HIM OPPORTUNITIES AND SOCIAL CONNECTIONS THAT FACILITATED HIS EARLY CAREER. AFTER STUDYING FORESTRY IN EUROPE, PINCHOT RETURNED HOME CONVINCED THAT THE FEDERAL GOVERNMENT WAS BEST SUITED TO IMPLEMENT LAND MANAGEMENT POLICIES BASED ON SCIENTIFIC RESEARCH OF AMERICAN FIELD CONDITIONS. HE ALSO RECOGNIZED THE NEED FOR A PROFESSIONAL ORGANIZATION AND EDUCATIONAL INSTITUTION TO SUPPORT A FEDERAL FOREST SERVICE. HE WANTED EACH OF THOSE TO HAVE A STRONG SENSE OF PROFESSION-ALISM AND PURPOSE—CHARACTERISTICS THAT QUICKLY BECAME THE HALLMARK OF THE FOREST SERVICE.

Pinchot Family Fortunes

The evolution of nineteenth-century American attitudes toward the land may be seen in three generations of the Pinchot family: Gifford's father, James; Gifford's grandfather, Cyrille Constantine Désiré Pinchot; and his great-grandfather, Constantine Pinchot. Supporters of Napoleon Bonaparte, Cyrille and Constantine had fled France after Napoleon's defeat at Waterloo. Within three years of arriving in the little town of Milford in northeastern Pennsylvania's Pike County, they had become successful dry goods merchants. They made most of their fortune, however, from land speculation and lumbering in eastern Pennsylvania, as well as in other parts of the country. A decade after arriving, the Pinchots were among the largest landholders in Pike County.

Like other lumbermen of his day, Cyrille maximized profits by quickly clearcutting the land and then selling it off for agricultural use before having to pay taxes. The lumbermen tied the logs together into rafts and floated them downriver to sawmills and markets in the ports of New York, Baltimore, and Philadelphia. Unstable markets meant that in some years, the lumber sold itself because of high demand, but in others, Pinchot and his business partner practically gave it away. Regardless of how much money came in, Pinchot reinvested in more timber stands, and the cycle repeated itself.[79]

In addition to savvy business practices, Cyrille taught James to use associates to his advantage, a lesson he later passed to his own children. In 1849, at age eighteen, James contacted a former teacher who worked for the Erie Railroad Company about selling lumber to the Erie, thus gaining access to a new, more stable market. Three years later, his former teacher helped with

the sale of Pinchot farmlands in northeastern Pennsylvania to families migrating to the area—families that would in all likelihood shop at the Pinchots' Milford store.[80]

When James struck the land sale deal, he had been living in New York City for two years. Shortly after his arrival there in 1850, he began developing an extensive social network along with numerous lines of business. He made his own fortune by selling wallpaper and other interior furnishings to the new office buildings and grand hotels going up in antebellum New York. His wealth permitted him to indulge his philanthropic impulses, such as funding contemporary American landscape art and artists, among whom were Sanford Gifford, whom he befriended, and others working at the famed Tenth Street Studio Building. The building itself was designed by one of its occupants, architect Richard Morris Hunt, who later designed Grey Towers in Milford for James and the Biltmore Estate for George Vanderbilt in North Carolina.[81]

While in New York, James married Mary Eno, who also came from wealth.[81] Mary gave birth to a son in August 1865, whom they named after Sanford Gifford. James and Mary raised Gifford and his brother, Amos, and sister, Antoinette, in privilege but with purpose. From a young age, Gifford's father imbued him with a sense of mission, perhaps one that compensated for the family's land speculation ventures.[83] The family's wealth left Gifford financially independent and above reproach when he entered government employment, and also meant that Gifford need not rely on a meager government salary for a living.

James purchased several works by Sanford Gifford and lent them for exhibitions in the United States and Europe. When not on exhibition, Gifford's *Hunter Mountain, Twilight* (1866) hung in

Pinchot homes in New York, Milford, and later Washington. The image, like others of the Hudson River School, juxtaposed a transformed land—a clearing with tree stumps to show human impact in the foreground—with a distant idyllic background bathed in warm light. Paintings by the Hudson River artists often documented vistas undergoing transformation and reflected a nostalgic yearning for the countryside before its development. Edgar Brannon, former director of Grey Towers National Historic Landmark, has observed that *Twilight* and other paintings like it that the Pinchots owned show that "the price we pay for civilization in part is the loss of nature," and that they illustrate "many of the paradoxes of conservation—that to live on this planet we must use its resources. But it's how we use the resources that's really the challenge."[84] The Forest Service has grappled with that challenge since its inception.

There is little doubt that living with such imagery in his home, as well as traveling to some of the locations depicted in the paintings, had an impact on Gifford Pinchot in his youth. Pinchot biographer Char Miller has noted the similarity of artists and foresters as seen through the impact of art on Gifford Pinchot:

> He had a painterly eye without the training of an artist. But that's actually not surprising because foresters have to have that eye also. They have to think out how forests grow. They have to understand how best to both manipulate and at times preserve these landscapes. The ax is both a tool to cut down and also to shape, and what else does an artist do but reshape landscapes on a canvas? In a sense, Pinchot was using this tool like an artist uses a paintbrush.[85]

Grey Towers, the Pinchots' summer estate situated on more than one thousand acres in Milford, Pennsylvania, was completed in 1886. James Pinchot built the house on clearcut land, which he rehabilitated with input from Gifford. (Grey Towers NHS)

Sanford Gifford's "Hunter Mountain, Twilight" (1866), when not on loan for exhibition, was prominently displayed in the Pinchot homes, including their mansion in Washington, D.C. James Pinchot collected several other Hudson River School images of cutover land, which had a notable impact on his son's thinking.
(USDA Forest Service)

Pinchot was not alone in possessing this skill. By the 1960s, in fact, the Forest Service had landscape architects shaping the design of clearcut areas to be more aesthetically appealing.

By the time Gifford entered Yale College in 1885 at age twenty, James had determined that his son would train for a career in forestry, even though forestry was not practiced in the United States and there were no forestry schools in North America. Gifford excitedly responded to his father's suggestion, "How would you like to be a forester?" by taking relevant science classes and reading everything on forestry and forests he could find.[86] It was an odd career choice for someone of his patrician background.

While at Yale, though, Pinchot was far from focused on his future profession. The tall, spare lad participated in several sports and was tapped to join Yale's prestigious secret society, Skull and Bones. Through extracurricular activities, he made lifelong friendships at Yale with men he called on repeatedly to join him in his life's work.[87] The skills and intelligence of several former classmates whom he later recruited helped establish the Forest Service and gave the agency its momentum in the early years.[88]

Studying Forestry in Europe

Shortly after graduating, Pinchot went to Europe to study forestry at the urging of Charles Sargent and Bernhard Fernow. Pinchot naively thought that he need only purchase a few books, visit an exhibit on forestry in Paris, and return home to work for Fernow as his assistant chief.[89] Fernow, constantly thwarted by Congress in his attempts to bring about forest preservation, let alone forest management, was pessimistic about a career in forestry. He advised Pinchot not to make forestry his primary profession, but rather to make it secondary to landscape gardening or some other profession that drew on the same sciences. Sargent and other forestry advocates agreed.[90]

After arriving overseas, Pinchot met the German-born forester Sir Dietrich Brandis. The German expert on silviculture had introduced German forestry techniques into Burma and British India in 1856, but now taught forestry at Cooper's Hill in Britain. He had been corresponding with Sargent, Franklin Hough, and others in the United States, closely monitoring the development of American forestry from afar.

Brandis encouraged Pinchot to enroll immediately at the French forest school at Nancy, gave him the necessary letters of introduction, and arranged several month-long apprenticeships for him with other leading European foresters.[91] He gladly became Pinchot's mentor, advising and teaching him about forestry and how best to bring it to the United States. Pinchot in turn sought his professional advice until Brandis's death, in 1907.[92]

Every prominent forester Pinchot met in Europe had a different opinion on what to do about what Pinchot called "our insane forest policy."[93] No one favored his working for a government that had no forests under active management. Instead, most suggested working as a private forester for wealthy patrons.[94] Encouraged during his time abroad, Pinchot became less apprehensive about a future in forestry, and the pessimism that Fernow had inspired faded.[95]

Pinchot learned much more than just the methods of his new profession during his European stay. Though he studied forestry in France and Germany, Pinchot's time with Swiss foresters proved most instructive in learning to apply forestry in a democracy. He noted that Swiss foresters worked within the nation's republican traditions and believed that their willingness to use a variety of forestry management techniques presented the best model for the United States.[96] He met Forstmeister Ulrich Meister, who was in charge of Zurich's Sihlwald forest, head of the city's liberal party, president of the largest Swiss newspaper, and a brigadier general in the Swiss Army—and he was writing a book. Pinchot also watched in amazement as Professor Elias Landoldt took shovel in hand and muddied his clothes while demonstrating to his students the

In 1890, Pinchot joined Sir Dietrich Brandis and his students on their excursions to German and Swiss forests. The field trips gave Pinchot the opportunity to discuss the American forestry movement with some of Europe's leading foresters, and for Brandis to assess Pinchot's forestry skills. Before returning home, Pinchot wrote, "What I should be as a forester without Doctor Brandis makes me tremble." Brandis is in the center holding the staff, and Pinchot is to his left, facing him. (Grey Towers NHS)

proper technique of tree planting.[97] Pinchot later emulated him by doing the same hard fieldwork as his rangers, often outshooting and outriding them as well.[98]

Pinchot's experiences in all three countries taught him that European silvicultural techniques could not be transplanted wholesale to his native country.[99] By the time Pinchot arrived in France, European foresters had been practicing even-aged management—clearcutting followed by artificial regeneration—for decades. To this day, foresters use clearcutting and replanting to produce predictable annual timber harvests and profits. In the United States, these methods are also used to grow shade-intolerant species that do not regenerate well unless a site is cleared.

Pinchot believed that even-aged management created sterile, uniform-looking forests that held little aesthetic appeal for Americans. European methods were "chiefly valuable as a sort of guide in the study of new conditions and the devising of new methods," he informed his father.[100] Instead, Pinchot would have "to conciliate the doctors, sportsmen, and practical men who look to the money return" to sell Americans on forestry. He had to walk a fine line between the aesthetic desires of the preservationists and the pragmatism required by forestry while meeting the demand for a continuous supply of timber. He wanted to adapt European methods to create an acceptable style of forestry in the United States—one that paid "a respectable income" yet left the forests "as picturesque as though left wholly alone."[101] Ultimately, he declared, his way would prevail: "I see no reason why our Forestry system should not be as unlike and superior in the end to the European as our Agricultural methods are."[102]

Pinchot advocated selection methods of silviculture, later called uneven-aged management, in which trees ready for harvest are removed while others are left behind for regeneration and future harvests. These methods are often used with species that need or tolerate full shade to sprout, and they reduce erosion; they are different, however, from "high grading," taking only the most valuable individual trees and leaving behind poor specimens to regenerate—a problem that American foresters continually encountered in the early years of forest management. Both clearcutting and uneven-aged management are intended to maintain a sustained yield or even produce more timber than is cut annually. Of course, the misapplication of any system can be problematic.[103]

Eager to return home, Pinchot left forestry school halfway through the program and before gaining a thorough knowledge of the sciences underlying forestry—he later lamented being "no more than half-trained"—and departed Europe with the advice of the Swiss foresters Lamboldt and Meister ringing in his ears.[104] He must get woodlands under active management and demonstrate forestry's viability in America, and he must do so immediately "before acting confessedly as a missionary" to bring forestry to America.[105]

Pinchot and the "Cradle of Forestry"

In late 1891, Pinchot rejected an offer from Fernow to work in the Division of Forestry and instead accepted George Vanderbilt's offer to initiate forest management on more than seven thousand acres at his sprawling Biltmore Estate outside Asheville, North Carolina. By 1895, under Pinchot's guidance, Vanderbilt had enlarged his holdings to some one hundred thousand acres. The forestry work in this private forest eventually earned the area the designation "Cradle of Forestry," even though forest management in the Adirondacks preceded Pinchot's work.[106] Pinchot took the Biltmore position with barely a year's schooling in France, only scant knowledge of the forests and trees of his native country, and little practical experience as a forester. Yet his unbridled enthusiasm and determination to justify the faith of Brandis, his father, and his other supporters compelled him to take the job and show that forestry could pay and would work in the United States.

As Pinchot would throughout his career, he drew upon available resources and proved himself adept as a manager of circumstances as well as people. He consulted with Sargent and with Vanderbilt's landscape architect, Frederick Law Olmsted, who knew Pinchot's father and had recommended the young man to Vanderbilt. He sent Brandis detailed accounts of the work, and Brandis followed the progress on a map Pinchot had provided. With his father, he discussed personnel matters and labor and land negotiations at Biltmore and even asked him to visit and help with various business affairs.[107] When his assistants wanted to know more about silviculture, Pinchot began lecturing to them in the evenings at what he half-jokingly called his "night school of forestry." He assigned readings and then discussed them in the next meeting. He also established

The Biltmore Estate became known as the Cradle of Forestry because of Pinchot's work and Carl Schenck's Biltmore Forest School. Pinchot made his reputation as George Vanderbilt's estate forester, which garnered him forestry consulting work for other wealthy patrons, including Vanderbilt's brother-in-law. Pinchot's consulting partner, Henry Graves, took this photo (facing page) while they were working for William C. Whitney, former secretary of the Navy, in the Adirondacks. (USDA Forest Service)

a Biltmore baseball team as a way to encourage esprit de corps.[108] In later years, summer students at the Yale Forestry School in Milford and other camps would field teams for similar reasons.

Pinchot's efforts in the Biltmore forests met with mixed results. The land had been cut over or burned for years by the locals, and only inferior trees were left behind for regeneration. Though Pinchot boasted that forestry turned a profit for Vanderbilt, he actually posted a loss on timber sales and showed a profit only by omitting his salary from the business ledger that first year. Nonetheless, he succeeded in generating interest in the viability of forestry while making a name for himself. Biltmore Forest served as a practical laboratory where visitors could see theory in action. Pinchot publicized his endeavors by exhibiting at the Columbian Exposition in Chicago and publishing a booklet on his work in 1893. Press coverage attracted young men interested in learning more about forestry or, at the very least, in pursuing a job that took them outdoors. Henry S. Graves, one of Pinchot's classmates from Yale, came to Biltmore. Won over to the cause by his charismatic friend, Graves studied under Charles Sargent at the Arnold Arboretum in preparation for studying forestry under Dietrich Brandis before returning home to become Pinchot's stalwart lieutenant.

Buoyed by his success, Pinchot exercised an option in his contract to do consulting work elsewhere while remaining in charge of forestry at Biltmore. Carl Schenck, a recent doctoral student of Brandis, emigrated from Prussia to take over day-to-day operations. With Pinchot's and Vanderbilt's blessings, Schenck established the Biltmore Forest School in 1898, the first forestry school in the United States. The school, which offered a one-year practical course in forestry aimed at those entering private forestry or the lumbering business, closed in 1913: it

could not compete with forestry schools that granted college degrees. Long before it closed, though, Pinchot and Schenck had parted ways because of opposing views on private lumber industry. Schenck had support from industry, while Pinchot blamed private industry's destructive clearcutting practices for creating the timber famine in the first place. As long as Pinchot was around, private industry remained a target of criticism.[109]

In 1893, Pinchot opened an office in Manhattan in the newly constructed United Charities Building, a gathering place for like-minded leaders of social politics.[110] He quickly found consulting work at Nehasane Park in the Adirondacks, a private reserve owned by George Vanderbilt's brother-in-law,[111] and undertook forestry management jobs for other wealthy patrons in the Adirondacks. With Graves, he published *The Adirondack Spruce*, the first American book to give a detailed plan for the management of a particular forest, in 1898.[112] Another job in 1894 led to a pioneering study of white pine in parts of Pennsylvania and New York by Graves under Pinchot's supervision. That project, partially funded by Pinchot's father, aimed to hasten, in Pinchot's words, "the general introduction of right methods of forest management," meaning selection methods. Only their use, he argued, could save American forests.[113]

A New Chief in Town

Having demonstrated the viability, if not the profitability, of forest management on private land, Pinchot turned his sights to bringing it to federal lands. His participation on the National Forest Commission and work as a confidential forest agent for the Department of Interior, along with his presence and outspokenness at pertinent meetings, made him the obvious candidate to succeed Bernhard Fernow in 1898 as chief of the Department of Agriculture's Division of Forestry. The problem was he did not want the job.

Pinchot's hesitation was understandable. The position in some ways would be a step backward for his career. He had a healthy consulting business and a growing national reputation. Furthermore, his wealth allowed him the freedom to pick and choose what he did. In contrast to Pinchot's rising career, the division was faltering, and Congress had asked the Department of Agriculture to justify continuing the division's funding.[114]

Charles Walcott of the U.S. Geological Survey and Secretary of Agriculture James Wilson, among others, believed that only Pinchot could save the division and slowly broke down his resistance. Secretary Wilson told him he could run the division as he saw fit, appoint his own assistants, do the kind of work he wanted, and not fear interference from him. Henry Graves promised to join him as assistant chief. After further consultation with others in the forestry movement, Pinchot accepted appointment as chief of the Division of Forestry. At age thirty-two, he seized the "chance of a lifetime."[115]

Pinchot took office on Friday, July 1, 1898, the first day of the fiscal year, and immediately swung into action. He worked all that day and the next, recording that he transferred one secretary out of his office and was making arrangements to bring in another.[116] His diary from the first few weeks recounts meetings both in Washington and elsewhere about planting experiments and forest reserve boundary lines. On his fifth day, Pinchot, eager to erase the negative perceptions of the Division of Forestry, asked Secretary Wilson to give him the title "Forester." In Washington, Pinchot noted nearly fifty years later, chiefs of divisions were many but "Foresters were not."[117] The change helped further delineate the differences between predecessor and successor and differentiate forestry in the public's perception as a unique science, different from horticulture and botany. (The title was changed back to "Chief" in 1935.)

Pinchot's small staff soon began expanding to meet responsibilities that seemed to increase almost exponentially. Just over three months after taking charge, Pinchot issued Circular 21, which offered technical advice to lumber companies if they paid the expenses of the department's agents.[118] The Kirby Company of Texas sought a management plan for its 1.2 million acres of pine forests, providing work for nearly fifty of Pinchot's employees. The expected response was quickly exceeded, with 123 applications for assistance from thirty-five states involving more than 1.5 million acres of private land. Pinchot justified the support to private landowners as being in the public interest—lumbering was the nation's fourth-largest industry and vital to the national economy.

Circular 21 was the first of many public-private cooperative agreements for the agency and the first substantial effort by the agency to educate private industry about the benefits of selection methods, conservative lumbering, and sustained yield. Given that the Division of Forestry's primary purpose was still providing information, this became a useful outlet for educating both

private industry and the American public about scientific forestry, and for giving men some practical training. As responsibility for public lands increased in the first years of the Forest Service, agency resources had to be reallocated, forcing the discontinuation of this successful program in 1909.

Cooperative agreements served Pinchot's needs in other ways. Because the division did not control the forest reserves, Pinchot needed permission from the Department of Interior to conduct forest management on the reserves. By 1901, the secretaries of Interior and Agriculture had forged a formal agreement. Interior would continue to patrol the reserves, enforce laws, and handle the paperwork; Pinchot's foresters in Agriculture would examine the reserves, make all technical decisions, and administer whatever plans they developed. Funding for these services came out of Agriculture's budget, and Pinchot reported on the reserves directly to Secretary of Interior Ethan A. Hitchcock. Secretary Hitchcock proclaimed it "the best solution of the problem which can be reached this year." At one point, Pinchot bristled under the system, until President Theodore Roosevelt reminded him that the situation was "not expected to be permanent."[119]

In the meantime, Pinchot made the most of the situation. He outlined for Hitchcock the "principles and practices…that should govern the administration of the National Forest Reserves." He argued for greater managerial flexibility to better respond to the diverse conditions found on the reserves and for courting those living on or near the reserves to help fight and prevent forest fires.[120] Hitchcock appears to have taken these ideas and turned them over to the newly created Division R (Reserves) within Interior's General Land Office in 1899.

Division R also bore Pinchot's imprint. Pinchot persuaded Filibert Roth to leave his teaching position at the New York State College of Forestry at Cornell and take charge of this new division

A capable administrator who willingly put in long hours in the office, Chief Pinchot advocated that the Washington men get in the field as much as possible. A visit from Chief Pinchot (center in above left photo) to a remote district, such as this one to the Yellowstone Timberland Forest Reserve in 1906, did much to boost morale in the early days. Pinchot also spent time with his fellow conservationists in the field (above), including Bureau of Reclamation chief, Frederick Newell (center). (USDA Forest Service)

Under the leadership of Filibert Roth (left), the Interior Department's Division R (for Reserve) struggled to implement forest management. Roth grew frustrated and quit after only fifteen months. H. H. Jones is on the right. (Library of Congress)

and sent a handful of his men to staff it. Pinchot appears to have written a "use book" in 1902, called the *Forest Reserve Manual for the Information and Use of Forest Officers*, though he credited E. T. Allen with its authorship.[121]

But Roth and his men faced a difficult situation. Division R was the third agency involved in managing the forest reserves, along with Pinchot's Division of Forestry and the Division of Geography and Forestry in Interior, and its situation within Interior associated it with the General Land Office's long history of political shenanigans, ineptitude, and corruption. Those who left Agriculture to work for Roth fretted that they were losing seniority while absent from the department. The men hired as rangers by Interior got the jobs mostly because they lived near the forest reserves they patrolled and knew the land and local customs. Some were political hacks appointed by their U.S. senators. Interior had gambled that local men would have less trouble enforcing regulations regarding grazing, trespassing, and illegal cutting; it had not envisioned rangers who would fail to fight fires or sleep on the job.[122] Roth's frustrations led to his resignation after just fifteen months and only reinforced Pinchot's belief in the need to transfer the reserves to his agency.

Transferring the Forest Reserves

Part of the publicity effort Pinchot launched after taking office in 1898 involved rallying support for the transfer of the reserves. Privately, Pinchot distrusted big business because he believed business leaders plundered in the name of profits and would not cease their destructive ways without government interference or regulation. But to secure the transfer, he had to court business. Circular 21 was part of that plan, as was his support of lumber tariffs. If lumber prices remained at profitable levels through tariffs and other means, it became more financially attractive to lumbermen to turn away from cut-and-run lumbering and embrace sustained-yield management practices.

Ranger H. C. Tuttle (left) and Than Wilkerson constructed the first ranger station in 1899 on the Bitterroot National Forest, Montana. In 1904, the site reverted to private ownership under a mining claim but was later purchased and returned to the agency. (USDA Forest Service)

Pinchot also sought presidential support. As early as 1897, during the fight over the Organic Act, President William McKinley had expressed his strong support for the reserves. McKinley welcomed the transfer idea when Pinchot's friend, Secretary of State Elihu Root, introduced the topic at a cabinet meeting in 1899. In early 1900, McKinley told Secretary Wilson that although he backed the transfer, in the face of stiff opposition, Wilson should not expect any measure to pass.[123] The following spring, the now-thriving forestry division was elevated to bureau status, but it did not receive the reserves to go with the name change.

President McKinley's assassination in September 1901 brought conservation-minded Theodore Roosevelt to office. A naturalist who had done some ranching in the Dakota Territory in the

William Kreutzer, the first forest ranger, served for more than forty years in Colorado. The uniform design was based on the military uniforms of the day, a practice that continued until the mid-1930s.

(USDA Forest Service)

1880s, Roosevelt had seen first-hand many of the problems conservationists had to handle. Pinchot and Roosevelt had known each other for years. Roosevelt was instrumental in Pinchot's admission to the Boone and Crockett Club, and while governor of New York, he had consulted with Pinchot about conservation and forestry. The two men saw eye to eye on how the executive branch should operate and were reluctant to wait for Congress to authorize their every action: within the bounds of the law, they thought it better to beg forgiveness than ask permission.

Three months after becoming president, Roosevelt made good on a promise to Pinchot and called for the transfer of the reserves to Agriculture, along with funding for reclamation as part of the broader federal effort to "fully regulate and conserve the waters of the arid region."[124] Perceived as a radical by the conservative wing of the Republican Party, Roosevelt knew he had to build his power base and demonstrate that he supported big business. Irrigation for the semiarid West indirectly got him the conservatives' support without much controversy. Though few opposed the idea, several eastern senators resisted initial efforts because their western counterparts wanted the federal government to pay for the reservoirs but turn control over to the states. Roosevelt worked with Representative Francis G. Newlands of Nevada, a Democrat, to ensure that his reclamation bill would not fall victim to the states' rights struggle. The Newlands Reclamation Act of 1902 represented the kind of national control over resource policy favored by the Roosevelt administration and foreshadowed what he wanted for the forest reserves.[125]

Political reality, though, prevented Roosevelt from pressing Capitol Hill too hard on the forest transfer issue. Despite pleasing western congressmen with the Reclamation Act in 1902, Roosevelt faced their continued opposition to the forest reserves during his first administration. When numerous bills calling for the forest transfer died in Congress, Pinchot tried other tactics. At his behest, Roosevelt appointed two commissions—the Committee on the Organization of Government Scientific Work and the Public Lands Commission—and placed Pinchot on both. Not surprisingly, the commissions came out in support of the transfer as a way of promoting governmental efficiency and cost savings.

In his annual messages to Congress, Roosevelt used language Pinchot had prepared stating the reasons for the transfer, but it was his election in 1904 that cleared the path for a final push. An enormously popular president, Roosevelt spent some of his political cache on Pinchot's behalf. In January 1905, Pinchot organized the American Forest Congress in Washington to help drive through the transfer bill. He named Roosevelt honorary president of the congress, and his immediate boss, Secretary Wilson, president. Scientific, civic, and political leaders delivered speeches, several of which were prepared by Pinchot and his publicity officer, Herbert Smith, that drew attention to the benefits of forestry and the need for sound government policy, which naturally included bringing the forests to the foresters.

In 2005, delegates from public and private organizations gathered in Washington for the Centennial Forest Congress to celebrate and to discuss the Forest Service's past and future. (Tom Iraci – USDA Forest Service)

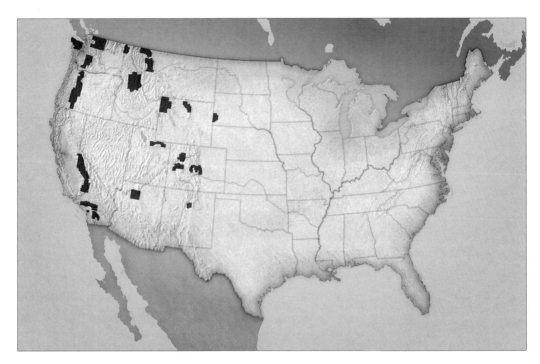

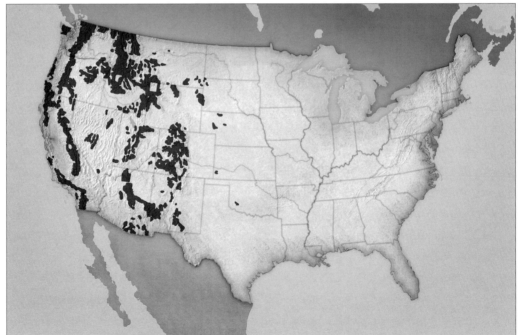

Congress quickly responded and presented the president with a transfer bill. On February 1, 1905, Roosevelt signed legislation authorizing the immediate transfer of the forests to the Department of Agriculture. Additionally, the Forest Service was permitted to retain all monies received from the sale of products grown in the government's forests, along with all fees for use

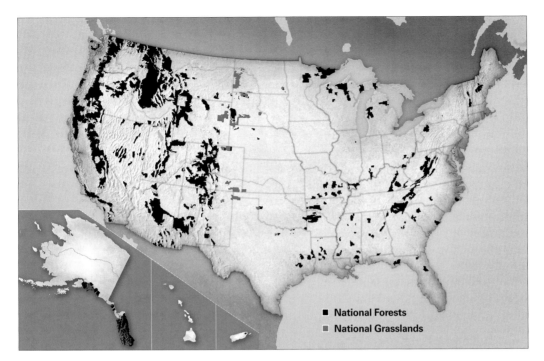

The Forest Service now manages nearly 193 million acres, including the national grasslands and one national forest in Puerto Rico. (Judy Dersh and Ruth Williams, USDA Forest Service)

■ National Forests
■ National Grasslands

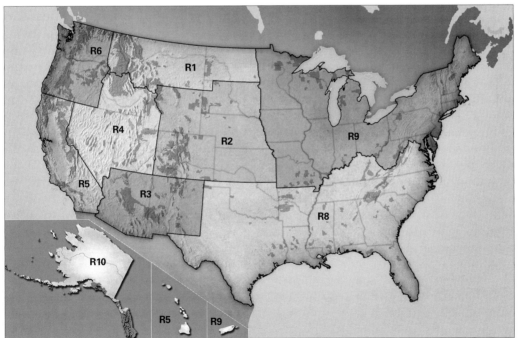

Initially, the national forests were divided into six districts (later called regions). With the establishment of eastern national forests following the Weeks Act of 1911, the number increased to ten. In 1966, Region 7 was combined with Region 9. (Judy Dersh and Ruth Williams, USDA Forest Service)

of the forests.[126] The transfer consolidated the various forestry programs of the General Land Office, the Geological Survey, and the Bureau of Forestry under one unit. One month later, the president signed a bill renaming the bureau as the Forest Service, effective July 1, 1905.

The Forest Service Badge

After a contest failed to turn up a satisfactory badge design for the new Forest Service, one emerged from an effort to prepare guidelines for a new competition. E. T. Allen, a contest judge along with Gifford Pinchot and Overton Price, insisted on a shield design that displayed the letters "U.S." to ensure quick public recognition of authority and also suggest public defense as a forestry objective. W. C. Hodge, who was watching Allen trace the Union Pacific Railroad shield, sketched a coniferous tree on a cigarette paper and laid it between the two letters Allen had drawn. They added "Forest Service" and "Department of Agriculture" to complete the design, and the contest was called off. In one hundred years, it is the only part of the Forest Service uniform that has not changed.

Source: "The Shield and the Tree," American Forests and Forest Life (July 1930): 392.

Two years after the transfer, the forest reserves were renamed national forests. Calling the agency the Forest Service instead of the Bureau of Forestry, Pinchot believed, signaled that federal foresters were there to serve the public, and that the land would continue to serve the needs of local people. And "reserves" connoted unavailability, or land reserved from use; "national forests" implied that the land belonged to the nation and its people.

Pinchot laid out the principles to be followed in administering the forests in a letter written for the Secretary of Agriculture's signature, which Secretary Wilson signed the same day Roosevelt signed the transfer bill. Those principles are still apparent in the Forest Service's fundamental policies. The letter directed that "all land is to be devoted to its most productive use for the permanent good of the whole people, and not for the temporary benefit of individuals or companies." Use of the land was not contrary to conservation. Furthermore, where conflict over who should benefit arose, "the question will always be decided from the standpoint of the greatest good of the greatest number in the long run."[127] With the transfer act, the foresters finally got their forests, and from Wilson's letter, they got their marching orders. Pinchot and his men were eager to begin but far from prepared.

Overnight, the Forest Service had sixty-three million acres under its control. Within months of the transfer, that figure rose to more than eighty-five million acres, and by July 1, 1906, it was nearly 107 million acres. The field force increased from 379 rangers, 87 guards, and 5 forest assistants in July 1905 to 511 rangers, 247 guards, 18 forest assistants, and 30 laborers one year later. The National Forest System expanded so quickly that in 1906 Pinchot reported that more than two million acres of the 107 million were not under organized administration. Even with the addition of forty-four million acres, he had the entire National Forest System under administration by the end of the fiscal year by reorganizing the agency and further expanding the field force. It meant, however, that each field officer engaged in patrol covered on average 132,236 acres.[128]

Creating a Profession

Pinchot had laid the groundwork for managing the public's forests before he formally assumed their control in 1905. As at Biltmore, Pinchot began by gathering the best possible men around him to leave himself free to concentrate on the big picture. Henry Graves joined Pinchot in the Washington office as his chief assistant. Graves was one of eleven staff members under Pinchot set up in two rooms on the third floor and a small space in the attic of the Department of Agriculture building. On his desk Graves posted a big sign commanding, "Silence." With a dozen men crowded into the largest room, it was the only hope of getting anything done. Despite their cramped quarters Pinchot and his staff enthusiastically took up their work.[129]

Circular 21 had increased Pinchot's need for trained men. Meanwhile, Pinchot and Graves had come to distrust what was being taught under Fernow at Cornell. They also differed with

Henry Graves left the Division of Forestry to start the Yale Forest School in 1900, where he taught for nearly three decades. Graves (back row, second from left) counted three future chiefs of the Forest Service—William Greeley, Robert Stuart, and Ferdinand Silcox (second row, second from right)—among his students. (USDA Forest Service)

Carl Schenck at the Biltmore Forest School. A forestry school at their alma mater, Pinchot and Graves decided, would train American foresters in American ways to work in American forests.[130] If the Pinchots would put up the money, Graves said, he would go to New Haven to head up the school. Early in 1900, after lengthy discussions, the Pinchot family gave $150,000 (a sum they later doubled) to Yale University to endow the Yale Forest School. Graves and James Toumey, who had given up his teaching position at the University of Arizona only a year before to lead the Division of Forestry's Section of Economic Tree Planting, left for Yale to prepare for the school's opening that fall.

The Pinchot family made seventeen hundred acres surrounding Grey Towers outside Milford, Pennsylvania, available for a field camp for the school. In conjunction with the summer program for the incoming class, a separate general forestry instruction program suitable for owners of woodlands, forest rangers, and schoolteachers was also offered.[131] Unusual for the times, the general course was open to women, though they had to stay in town at a boarding house instead of in the camp itself. After the third year, women stopped attending, though it is not clear why, and Yale discontinued the general instruction program after 1908 because of a drop in applications.

James Pinchot meanwhile constructed buildings, including a small library and a classroom at the campsite, built a lecture hall in the town, and purchased surplus army tents to house the students. In 1903, he established the Milford Experimental Station, to be run by the school's dean, and provided office space for the station in the town's library, which the family had also built. The Milford site operated as Yale's primary fieldwork location for nearly twenty-six years and educated more than five hundred foresters. Yale closed the site when the land no longer met the students' fieldwork needs.[132]

The Pinchot family hosted Yale Forest School's summer camp at Grey Towers for a quarter of a century. Along with classroom training, field training on the Pinchot estate included surveying (note Grey Towers in the background above) and such pragmatic lessons as how to protect oneself if confronted by a bear. (Top photo, Grey Towers NHS; bottom photo courtesy of Manuscripts and Archives, Yale University Library)

If the Biltmore Estate and Pisgah Forest in North Carolina are the cradle of forestry, then Grey Towers in Pennsylvania must be considered its nursery. Here, under the delighted gaze of the Pinchot family, the Yale faculty nurtured and educated many of the first two generations of American foresters. For seven weeks every summer, young men from any number of colleges came for the summer course in forestry or began their two-year stint at Yale. At Grey Towers, they learned surveying on the front lawn of the estate and forest mensuration in its woods; skinny-dipped in the Sawkill Falls (much to the shock and amusement of the locals who came to watch); walked to and from town singing songs about faculty members; played against local baseball clubs and held tennis tournaments on the two grass courts in camp; dined with the Pinchots at least once a summer at the main house; gathered around a bonfire at night to listen to Gifford and other conservation leaders discuss what forestry meant to the nation's future—all the while forging the esprit de corps Pinchot had first learned from Brandis and had tried to instill in his employees at Biltmore.

Although Pinchot's plans to create an institute of conservation education fell short when he failed to get backing for a school of irrigation, Yale Forest School sent so many graduates to Washington that the Forest Service headquarters resembled a Yale alumni chapter. Rumors of favoritism toward the sons of Eli led a chief some years later to openly refute the charge,[133] but in fact, the first five chiefs of the Forest Service came from Yale.[134]

Shortly after the Yale Forest School opened in 1900, Pinchot and several employees established the professional organization known as the Society of American Foresters (SAF), which soon began publishing a professional journal, known today as the *Journal of Forestry*. Once again, Fernow had anticipated him by launching his own professional forestry organization and journal at Cornell the year before, but Pinchot's position and acumen enabled his organization and journal to quickly overwhelm the competition.

Pinchot was elected president of SAF and held its meetings in his Washington home on Rhode Island Avenue for the first several years. The meetings often featured guest speakers from science and politics to discuss forestry and other conservation matters. President Roosevelt spoke at one meeting about the importance of forestry to American civilization, a moment vividly recalled later by those present and a testament to the close friendship of the president and a lowly bureau chief. SAF gave forestry immediate standing among other emerging scientific professions, and Yale gave Pinchot a steady supply of highly motivated foresters ready to carry forth the message of conservation. Pinchot not only created the Forest Service, but through Yale Forest School and the Society of American Foresters, he gave birth to the profession of forestry in the United States.

The Four Levels of National Forest Offices

National level: The person who oversees the entire Forest Service from the Washington office is called the chief. The chief is a federal employee who reports to the Under Secretary for Natural Resources and Environment in the U.S. Department of Agriculture (USDA).

Region: There are nine regions, numbered 1 through 10 (Region 7, the Northeast, was combined with Region 9, the Midwest, in 1966). With the exception of Alaska (Region 10), each region comprises several states. Headed by the regional forester, the regional office staff coordinates the activities of its national forests, monitors its national forests to ensure quality operations, provides guidance for forest plans, and allocates budgets to the forests. Until 1930, regions were called districts.

National forests: There are 155 national forests and 20 national grasslands. Each national forest is managed by a forest supervisor and comprises several ranger districts; a district ranger manages each district. The headquarters of a national forest is the supervisor's office. This level coordinates activities of the forest's districts, allocates the budget, and provides technical support.

Ranger district: The district ranger and his or her staff is often the public's first point of contact with the Forest Service. There are more than six hundred ranger districts, each with a staff of ten to a hundred people. The districts range from fifty thousand acres to more than a million acres. On-the-ground activities include trail construction and maintenance, operation of campgrounds, and management of vegetation and wildlife habitat.

Source: Adapted from "About Us—Meet the Forest Service," accessed at http://www.fs. fed.us/aboutus/meetfs.shtml, 1 September 2004.

Hiring the Best and the Brightest

When Henry Graves left for Yale, Pinchot promoted Overton Price, who had trained under Carl Schenck, to associate chief. Price ran the office in Pinchot's absence so effectively that Pinchot was still praising him in his memoirs more than forty years later: "To say that Price was my right hand is a feeble understatement. He had more to do with the good organization and high efficiency of the Government forest work than ever I had. Most of the credit for it that came to me rightly belongs to him."[135] Price snapped out commands in succinct, business-like prose and answered questions with direct, one-paragraph responses. During Pinchot's four-month trip around the world in 1902, Price ran the agency without problems.[136]

Price also brought order to chaos. He formulated a reorganization plan, initially adopted in 1903, that redistributed staff responsibilities and relieved Pinchot of administrative minutiæ. Two years after the 1905 transfer, by which time the agency controlled more than 150 million acres, Pinchot submitted a plan to expand the three inspection districts into six (now called regions) and placed district foresters (now called regional foresters) in charge. E. T. Allen, William B. Greeley, Clyde Leavitt, Frederick E. Olmsted, Smith Riley, and Arthur Ringland were the first six district foresters.

The arrangement gave those in the field more autonomy to deal with problems, even down to the district ranger level, instead of inundating the Washington office with questions. Pinchot had decided on this decentralized approach even before the transfer: "In the matter of judgment concerning forest problems, the field man who knows the conditions, should rule." Rangers who had to answer to their neighbors behaved differently from those who could blame Washington for everything. It made them a part of the community they served—and it meant that local constituents received immediate responses.

Shortly after the transfer of the reserves, with the Washington office overwhelmed by its new administrative responsibilities, Pinchot charged Price with improving support to field personnel. Price was responsible for answering all requests for supplies, from ink and paper for district offices to firefighting tools, within twenty-four hours and received authorization to bring in men from the field to make that happen. To save money and deliver supplies faster to field personnel, the agency established a supply depot in Ogden, Utah. The only priority greater than improving field support, Pinchot believed, was creating new forest reserves.[137]

In 1900, Pinchot brought Albert Potter east from Arizona to handle overgrazing and soil erosion. A former cattleman who had begun raising sheep in 1896, Potter was secretary of the eastern division of the Arizona Wool Growers Association, arguably the best-organized livestock interest in the West at that time. When Pinchot went to Arizona to discuss range problems, Potter persuaded him to reconsider a ban on sheep grazing on the forest reserves and to render a decision after examining field conditions. On the recommendation of Pinchot, the Interior Department reversed its position. Anyone wishing to graze sheep had to obtain a free permit,

Overton Price

Albert Potter

George Woodruff

Herbert Smith

Architects of the early Forest Service: Overton Price effectively handled day-to-day operations as associate chief; Albert Potter was recruited by Pinchot to manage range and grazing issues; Herbert Smith oversaw public relations for three decades; and George Woodruff, with his deft legal skills, established important precedents for the agency. (USDA Forest Service)

which limited the number of animals, and set the duration and location of grazing in the reserves. Thoroughly impressed with the stockman, Pinchot decided that Potter was the best man to carry out the new grazing policy and invited him to join the Division of Forestry. Potter was the first westerner to hold a high post in the division.

Potter's position within the industry gave him instant credibility when discussing the need and right to impose fees and control access to public land. He became "the cornerstone upon which we built the entire structure of grazing control," Pinchot recalled.[138] Grazing was the major activity in the national forests until World War II. By the 1940s, grazing took place on 100 of the 152 national forests. Hiring the right person to set and implement policy was critical to the overall success of the agency's multiple-use philosophy, which under Pinchot quickly exceeded what the Organic Act set down as the purposes of forest reserves—protecting the forests, furnishing a timber supply, and offering watershed protection. Potter's struggle to establish the power of the federal government to maintain its rules and regulations was his most important contribution during his nineteen years of service.

The laws that Congress passed had to be interpreted so that the Forest Service would know its legal limits. To handle legal affairs, Pinchot brought in George W. Woodruff and Philip P. Wells, two friends from Yale. Woodruff lost little time in letting people know that the agency tolerated no missteps or fraud, and its own employees were not excepted. When Forest Supervisor Everett B. Thomas was accused of falsifying his accounts before the 1905 transfer, Woodruff worked hard for a conviction. By setting an example for other supervisors, Woodruff declared, "the moral influence of such a conviction throughout the Service will be very great." The conviction would lead others "to consider the ethics of their actions." Thomas received a three-year sentence and $7,000 fine. Pinchot never had to worry about his employees after that.[139] Potter's work and Woodruff's legal prowess alone would not convince the public or Congress of the need to transfer the forest reserves to the Department of Agriculture, so Pinchot took his case to the press. To handle the task, in 1901 Pinchot tapped yet another Yale classmate, Herbert

"Dol" Smith, to edit publications. Smith had been teaching English at Yale when Pinchot recruited him, and soon enough he was handling publicity and writing press releases for newspapers around the country. Pinchot believed that educating the public would ensure acceptance and continuation of agency policies and build support for the transfer of the reserves. Smith made that possible. He served as adviser to Pinchot and several other chiefs and drafted the agency's annual reports, along with speeches for Pinchot and other government officials. He remained with the Forest Service in that capacity for thirty-six years.

The success of Circular 21 and the dearth of forestry schools and foresters left Pinchot scrambling for staff, so he employed young men, most of whom were college students, to round out his office and field personnel. The student assistant program allowed Pinchot to meet staffing needs on a tight budget. The program generated publicity, fostered esprit de corps, and trained men quickly so that the agency could conduct its work efficiently. The program gave the men a taste of professional forestry, usually at its most unglamorous. They lived in lumber camps or in the field and were "expected to assist in whatever work is assigned," which generally involved surveying the reserves or cruising timber to determine how much was on a parcel of land. They earned $25 a month and had to pay their own travel expenses from the Washington office after orientation and back home after the season.[140]

The work was hard. Survey crewmen "developed legs of steel," recalled Arthur Ringland, a veteran of three seasons, as the men "struggled through dense rhododendron and laurel in an Appalachian cove, a cypress swamp in East Texas, or a spruce blowdown in Maine" while examining land. For Ringland, one of the first student assistants, the arduous work only served to inspire him to enter forestry, as it did so many others. "Experience as a student assistant with many field parties, the influence of my associates, and the inspiration of my chief Gifford Pinchot and his brilliant associate forester Overton W. Price, fortified my decision to follow the profession of forestry," he recalled.[141]

For Pinchot and the division, the program proved an unqualified success. Pinchot accepted twenty-eight student assistants in 1899 and by 1901 had nearly three hundred. These budding foresters—who were pointedly not "lumbermen"—had opportunity to explore the land as well as the profession. The students and their supervisors forged the camaraderie Pinchot had hoped for. Several participants, including Ringland, Coert DuBois, and Ferdinand A. Silcox, later rose to high positions in the Forest Service.

Participation in the program did not guarantee a job, however. Every applicant to the division—whether for student assistant or for fulltime work—had to sit for "a severe technical" civil service examination that allowed Pinchot to avoid accusations of favoritism.[142] When Pinchot received his appointment in 1898, he learned his job fell under the civil service rules and that he would have to pass an exam. Since no one else in the government was qualified, he drew up an exam for himself and was fully prepared to take it when President William McKinley waived the

requirement. Those working for Pinchot, including Henry Graves, his first associate chief, received no such waiver. Even he had to sit for an exam, one drafted and administered by his former classmate. The emphasis on hiring only qualified men reflected Pinchot's determination to keep forestry a profession, not a political plum.

Rangers and the *Use Book*

The agency needed men who could do many things competently. Consequently, the first two generations of foresters were generalists who lacked specialized training. In the field, rangers had to be jacks-of-all-trades as well as courteous and above reproach. The stringent requirements no doubt influenced several men who came over from the Interior Department with the transfer in 1905 to reconsider keeping the job. Those hired before the transfer were not required to have many skills beyond being familiar with the local area. To keep their jobs afterward, however, they had to follow a new set of rules and expectations. The *Use Book* declared, "Forest officers are agents of the people. They must answer all inquiries fully and cheerfully, and be at least as prompt and courteous in the conduct of Forest business as in private business. They must obey instructions and enforce the regulations for the protection of the Forests without fear or favor, and must not allow personal or temporary interests to weigh against the permanent good of the Forests."[143]

Student assistants provided critical help in the early days by aiding in timber surveys and other fieldwork at low wages. Although most student assistants worked in the West, Arthur Ringland (fourth from right) served in parties from Maine to Texas over three seasons. Ringland eventually became one of the first six district rangers.
(USDA Forest Service)

Even in the field, a ranger was expected to maintain a professional appearance. Because reconnaissance and survey parties spent weeks in the field at a time, field barbershops, such as this one on the Carson National Forest in 1911, helped the men keep up appearances. (USDA Forest Service)

The *Use Book* evolved from the forest officer manual issued by the Interior Department's Division R in early 1902, and many of its tenets carried over into the manual issued when the Forest Service came into existence on July 1, 1905. A mere 142 pages in length and small enough to slip easily into a ranger's shirt pocket, the *Use Book* laid out Forest Service policy in typical Pinchot style—concise, clear, and direct. Not surprisingly, as the agency's responsibilities grew, so did the size of the book. By 1907, it exceeded 240 pages, but the mission remained the same. "Forest reserves are for the purpose of preserving a perpetual supply of timber for home industries, preventing destruction of the forest cover which regulates the flow of streams, and protecting local residents from unfair competition in the use of forest and range. They are patrolled and protected, at Government expense, for the benefit of the Community and home builder."[144] Pinchot recognized that without local support, the National Forest System was doomed to fail. His support of the Forest Homestead Act in 1906, which granted permanent settlers up to 160 acres of agricultural land within forest reserve boundaries, was an extension of this policy.

Rangers had to demonstrate the ability to use an ax, shoot, ride, and throw a diamond hitch, a difficult knot used to lash freight on a mule or horse. In addition, they had to pass a written test. They also had to prepare a meal in the field—and eat it. A ranger had to be "thoroughly sound and able-bodied, capable of enduring hardships and of performing severe labor under trying conditions." He needed to know "something of land surveying, estimating and scaling timber, logging, land laws, mining, and the live-stock business." In addition, those

THE BOOK OF KNOWLEDGE

THE BOOK OF KNOWLEDGE

THE OLD USE BOOK

ADMINISTRATIVE GUIDE

THEN

V.C—.

NOW

in the territories of Arizona and New Mexico needed to "know enough Spanish to conduct Forest business with Mexicans."

Once hired, they were subjected to rigorous inspection and required to keep tidy and sanitary ranger stations as examples to campers. They were even issued standards to follow in preparing privy sites. Rangers could expect to work from sunup to sundown or later, seven days a week, and often alone for days on end. Pinchot wanted men with experience with the region and its people, not just books. A ranger was often the only person enforcing regulations on a reserve of several hundred thousand acres, not to mention the "only policeman, fish and game warden, coroner, disaster rescuer, and doctor. He settled disputes between cattle and sheepmen, organized and led fire fighting crews, built roads and trails, negotiated grazing and timber sales contracts, carried out reforestation and disease control projects, and ran surveys." The *Use Book* made clear that this was not work for weaklings or tenderfoots: "Invalids seeking light out-of-door employment need not apply."[145]

Like rangers, forest assistants also reported to the forest supervisor. Forest assistants—the technical men straight out of college—were hired to prepare maps and working plans but did a variety of jobs as needed. They were generalists who learned many of their skills on the job. They also had to quickly adjust their thinking to the vast western landscape. While training in the East, they worked on woodlots covering a few hundred to a few thousand acres; on western national forests, they were dealing in millions of acres and billions of board feet. Forest assistants soon learned that it helped community relations if they downplayed their "Eastern effete" college backgrounds and dressed like the locals, rather than being "rigged out from head to foot" in

Originally designed to fit in a ranger's shirt pocket so that he could carry it in the field, the *Use Book* grew in size with each passing year. By 1930, when this cartoon appeared in a book celebrating the Forest Service's twenty-fifth anniversary, the manual was thick enough to make a bookshelf groan under its weight or, in this case, collapse a table. (USDA Forest Service)

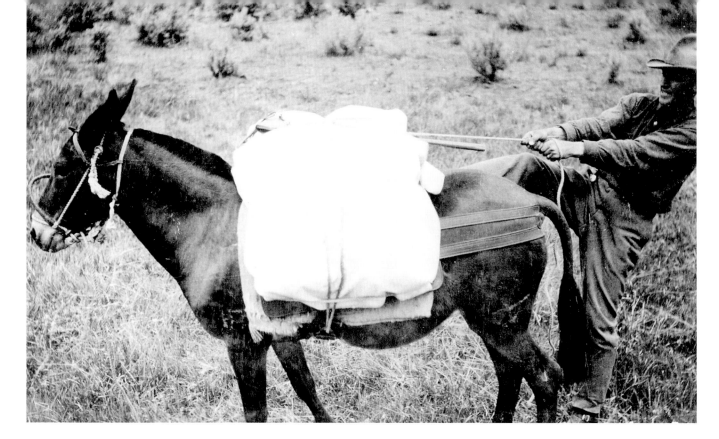

Part of the early ranger examination was demonstrating knowledge of packing supplies and equipment for safe transport. The key to success in loading a pack animal properly was the diamond hitch, as demonstrated here by a ranger in 1912. (USDA Forest Service)

Abercrombie and Fitch's high-priced sporting outfits.[146] One experienced staff member advised the Washington office to "get the Harvard rubbed off the students before they came in contact with the loggers."[147]

Defining Conservation

Pinchot wedded philosophy to politics—conservation with the Progressive movement—when he created the Forest Service. Although enabling legislation authorized the federal government to protect watersheds and timber, Pinchot placed those efforts in a broader conservation policy. Developed during discussions with close associates who were also government scientists and bureau chiefs, such as W J McGee and Frederick Newell, and with President Roosevelt's encouragement and support, the policy that Pinchot crafted drew inspiration from the ecological reporting of George Perkins Marsh and the utilitarian thinking of philosopher Jeremy Bentham, as well as the writings of the time. Pinchot sought, as he put it, "the use of the natural resources for the greatest good of the greatest number for the longest time."[148] This phrase has been the core principle of the Forest Service since the agency's creation in 1905: conserving the land for multiple uses and managing the land to support all of them as well as possible for sustainable use. As historian William Cronon has noted, "The issue of sustainability, in effect, gets imported into utilitarianism by this new principle."[149]

What has changed over the past one hundred years is how each phrase in the statement has been defined, and by whom—the Forest Service, other land management agencies, Congress, business, the public, preservation advocates, recreational users, and others. How do the agency and its numerous constituencies define "the greatest good"? Who or what constitutes "the greatest number"? How long is "the longest time"? Changing the definition of one phrase affects the other two. What the Forest Service has struggled with—and will always struggle with—is balancing all three. Pinchot's philosophy could not answer all questions regarding land management, but it offered an alternative to the thoughtless destructive policies of the past. It provided a starting point for a discussion about managing natural resources—a conversation the government and the public needed to have in 1905, and one that continues today.

Under Pinchot's conservation plan, scientific expertise would inform the questions of how to use the land efficiently to benefit all humans, not just a select few who used it for personal profit. With forestry as its cornerstone, conservation included irrigation and reclamation to save watersheds and waterways, and biology and botany to sustain the rangelands and game animals. The initial goal of forest conservation was simple: to protect watersheds and timber by regulating the ax. Political and economic reality meant managing the land for use—however use would be defined—and not absolute preservation. Pinchot favored commercial activity to gain and keep western congressional support. He would allow mining, grazing, and other extractive activities in the national forests to continue while providing protection for game animals, wildlife, and wilderness.

Though pockets of opposition toward the Forest Service remained in the West, public support for Pinchot's policies was growing. He soon began labeling his policy as conservation.[150] Before Pinchot, government agents charged with devising irrigation, forestry, animal preservation, and other policies usually worked independently of each other, resulting in overlapping jurisdictions and confounding the work of competing bureaus in the Agriculture and Interior departments.[151] After 1907, the Forest Service's efforts noticeably expanded beyond timber conservation to include all natural resources found on the national forests.[152] The Forest Service thus became a leader of the broader conservation movement.

When talking to his employees and forestry students about what conservation meant to the nation, Pinchot spoke of the greatest good in the loftiest terms. From his officers in the Washington office to the lowly student assistants in the field, he instilled in his employees a belief that they were selfless, heroic civil servants working for a cause greater than themselves. Sustainable forestry meant sustaining democracy itself. He succeeded where others had failed because he had the active support of the executive branch whose leaders understood that an emerging industrial economy needed a sustainable supply of renewable natural resources. Conservation provided the framework on which to build that industrial economy.

What has sometimes been lost over the decades is that for Pinchot, the greatest good meant not just development of the land for its natural resources, but also preservation of natural resources for aesthetic reasons. Indeed, he took up forestry because it offered time in the outdoors,

Gifford Pinchot's conservation plan included the rational development of waterpower sites. By leasing sites instead of selling them, the agency prevented corporations from establishing hydroelectric monopolies that owned waterways. (USDA Forest Service)

something Pinchot, an avid sportsman, cherished throughout his life. He favored wilderness areas and the creation of national monuments to preserve such splendors as the Grand Canyon. His diaries and other writings are full of awed descriptions of such places.[153]

One flaw in conservation as envisioned by Pinchot and implemented by Roosevelt was the very concept that gave conservation much-needed credibility at the outset—the use of nonpartisan bureaucrats and experts to formulate policy. Believing natural resources management to be too complex to be understood by the public and too important to be left to the whims of the democratic process, Roosevelt and Pinchot wanted to empower a permanent establishment to handle it in a nonpolitical and disinterested way. Pinchot replaced what he called "our insane forest policy," which was not a forest policy at all, with an apparatus for implementing one based on scientific management. He put his faith in government scientists because it was morally, scientifically, and economically the right thing to do, but in so doing, he isolated the agency from the public and Congress. Forest Service leaders would slowly lose touch with the constituents they served, leaving their agency vulnerable to increasing criticism after the 1950s.[154]

Edgar Brannon of the Forest Service described what Pinchot had accomplished in setting up the Forest Service this way: "What Pinchot gave the Forest Service is a set of moral values [and] an underlying belief system that transcend forests in themselves. Those values are that the natural resources of the world are the basis of our national well-being, and that that national well-being—the value of those resources—needs to be shared amongst everybody. That [the forest is] not—that it should not—and cannot be used for the benefit of the few, it has to be used for the benefit of the many."[155]

Peter Pinchot, like his grandfather Gifford a forester, summed up the first American forester's conservation this way: "His greatest contribution was coming up with what I think is really a new social contract about the relationship between people and nature. That social contract included the idea of benefiting all people, not simply the individual. The Jeffersonian contract was about individual liberty and individual rights. Gifford Pinchot's contract was about social benefits and about this generation and future generations."[156] That social contract has required the Forest Service to continually meet the changing needs of the public and the land.

Chapter Three

POWERS, POLICIES, AND FIRES

Buck Mountain Lookout on the Blue River Ranger District, 1960. (USDA Forest Service)

T HE ESTABLISHMENT OF THE FOREST SERVICE IN 1905 INDICATED THE FEDERAL GOVERNMENT'S COMMITMENT TO CONTROL LAND PRIVATIZATION ON THE FRONTIER. IT WAS ONE THING TO CREATE A LAND MANAGEMENT AGENCY TO MANAGE PUBLIC LANDS AND THEIR RESOURCES IN PERPETUITY; IT WAS QUITE ANOTHER FOR THE AGENCY TO HAVE THE LEGAL AUTHORITY TO MAKE THAT POSSIBLE. CONGRESS DID NOT GIVE THE FOREST SERVICE EXPLICIT AUTHORITY TO REGULATE GRAZING, THE MOST IMPORTANT COMMERCIAL ACTIVITY ON THE FOREST RESERVES, OR HUNTING OR OTHER ACTIVITIES. GIFFORD PINCHOT AND HIS FOREST SERVICE OFFICERS WOULD HAVE TO FIGURE OUT NOT ONLY WHICH ACTIVITIES TO REGULATE, BUT HOW TO DO SO IN PERPETUITY.

Grazing on the National Forests

Conflict between cattlemen and sheep grazers, which sometimes turned violent, had been raging for more than thirty years in the West when the General Land Office in the Interior Department took up administration of the forest reserves in 1898.[157] Interests concerned with watershed protection and irrigation wanted grazing banned from all forest reserves because of the damage it caused to groundcover. Gifford Pinchot of the Agriculture Department's Division of Forestry recognized, however, that because grazing was the primary commercial use of the forest reserves, the livestock industry was a powerful political force. He courted the livestock producers to gain their support for transferring the forest reserves to his agency.

The General Land Office reluctantly acquiesced to Pinchot and agreed to issue free grazing permits for cattle on all reserves in 1898 as a way to limit the number of animals on each reserve. The agency initially restricted sheep to forest reserves in Oregon and Washington because prevailing wisdom blamed sheep for the majority of range destruction. Naturalist John Muir summed up the general opinion about soil erosion and watershed protection when he referred to sheep as "hooved locusts." Theodore Roosevelt spoke for many cattlemen when he called sheep "bleating idiots" that ate the grass so close that cattle could not live on the same land.[158]

To formulate a permanent policy, the General Land Office also asked Department of Agriculture botanist Frederick V. Coville to provide scientific evidence. Coville's report on grazing in western Oregon recommended that sheep grazing in the region's forest reserves be regulated, too, but when the government reneged on the policy one year later in 1899, the powerful sheep industry appealed to Pinchot for help. Pinchot met with Albert Potter of the Arizona Wool Growers Association and examined western rangelands to review range conditions. The resulting compromise between grazing interests and the federal government allowed supervised sheep grazing on additional reserves.[159] Soon thereafter, Potter joined the agency. The Forest Service

Regulating grazing to protect forage and watersheds was one of the first responsibilities of the Forest Service. On the Wasatch National Forest in Utah, in 1914, (above) this was one way that sheep heading into the open range were counted. The benefits of grazing regulation can be seen on the Santa Fe National Forest (facing page), which in 1924 showed no injury to the forest. (USDA Forest Service – Forest History Society)

inherited Interior's permit system when the agency took charge of the forest reserves in 1905. Given Pinchot's emphasis on "the greatest good for the greatest number," grazing continued under Potter's direction because the Forest Service considered it a proper forest use and grazing fees a potential source of revenue. In 1906, the Forest Service announced nominal grazing fees to cover its administrative costs. When rangers explained the long-term benefits of conservative use of the forest to the larger cattle stockmen, most ceased their grumbling and came out in support of fees.[160] The Forest Service issued nearly eight thousand grazing permits in the first year after the transfer. The agency's fee-permit system slowly brought order to a chaotic situation.

The policy did not sit well with everyone, though, as cattlemen and sheepmen competed for permits and the fee rates drove smaller stockmen out of business.[161] When Fred Light turned his five hundred head of cattle loose to graze on Colorado's Holy Cross Forest Reserve to challenge Forest Service permit requirements in 1906, a ranger charged Light with trespassing. Light filed suit, and his appeal slowly wound its way to the U.S. Supreme Court. George W. Woodruff, Pinchot's first law officer, and Philip P. Wells, Woodruff's assistant and successor after Woodruff moved to Interior in 1907, used this and other cases to build legal precedents. By filing charges against selected violators in the courts of supportive judges, Woodruff and Wells constructed legal support for Forest Service policies. The Supreme Court upheld the lower court's injunction against Light on May 1, 1911.

Two days later, the Supreme Court upheld the Forest Service's right to charge grazing fees with the *Grimaud* decision. Pierre Grimaud had grazed his sheep on California's Sierra National

Forest without permission in an effort to challenge the validity of making a violation of Forest Service regulations a penal offense. Together, the *Grimaud* and *Light* cases upheld the constitutional right of Congress to set aside and reserve portions of the public domain as national forests, to prescribe penalties for violating regulations, to give the secretary of Agriculture the right to make rules and regulations for the national forests, and to impose fees for grazing permits. In short, these two cases established beyond a doubt the constitutionality of the national forests and the Forest Reserve Act of 1891 and the Organic Act.[162]

Thus, range management problems enabled Pinchot to set important precedents in public land management the agency could apply elsewhere. In formulating policy, Pinchot and the Forest Service worked to reach compromises that both sides could live with. When challenges to those policies went to court, the agency demonstrated the patience and resolve necessary to establish "an impressive corpus of common law to give substance to hard-won legislative battles."[163]

The Sagebrush Rebellion

Fred Light's case was a sign of the growing western resentment of federal control. Pinchot tried to assuage it by showing that he had local interests at heart. Pinchot strongly advocated for the Forest Homestead Act of 1906 because the law officially recognized local primacy.[164] That same year, Congress passed legislation that directed the return of ten percent of the proceeds from the sale of timber on the national forests to the counties in which the national forests were located. Two years later, Congress increased the percentage to twenty-five percent. The act

Czar Pinchot and His Cossack Rangers Administering the Forest Reserves

CRIMES BY REGULATIONS

1. THE OWNERS OF STOCK THAT STRAYS UPON FOREST RESERVES WILL BE FINED AND IMPRISONED.
2. NO STOCK CAN FEED ON THE PUBLIC RANGE WITHOUT PAYING A HEAD TAX.
3. EVERY PERSON WHO CUTS TIMBER ON FOREST RESERVES FOR ANY PURPOSE WHATEVER DOES SO AT HIS PERIL.
4. SETTLERS WITHIN FOREST RESERVES ARE UNDESIRABLE CITIZENS.
5. WILD GAME IS OF MORE IMPORTANCE THAN PROSPECTORS OR SETTLERS.
   GIFFORD PINCHOT. CHIEF FORESTER

STOCKMAN   IRRIGATIONIST   MINER   NEW SETTLER   PROSPECTOR

News Sept. 20-08.

As the government official most closely associated with developing and implementing many of the new public land regulations, Gifford Pinchot was a frequent target of western editorial cartoonists. (Courtesy of Denver Public Library)

undercut complaints that national forests would deprive counties of future property taxes and did not serve local interests. The long-term ramification was that it made the sale of national forest timber a matter of great economic importance to the affected communities, and it encouraged congressional representatives to demand higher and higher timber harvest levels from the Forest Service to support their constituents.

Pinchot's attempts at conciliation did little to slow the mounting opposition. Congressmen from all parts of the country joined in the chorus of criticism over President Roosevelt's expansion of executive power in general.[165] The Colorado legislature passed a special appropriations bill to defray the legal expenses Fred Light incurred while he fought the federal government and sent the state's attorney general to represent him before the Supreme Court. Western congressmen unsuccessfully tried to stop the Forest Service by limiting funding or restricting the agency's powers. Some conservative easterners, led by Speaker of the House Joe Cannon, lent support to the "Sagebrush

Rebellion," a backlash against federal control of western lands based on states' rights arguments.[166] Several western leaders, on both federal and state levels, demanded local instead of federal control. The off-year elections in 1906 and the need to show party unity to maintain control of Congress temporarily quelled the attacks from within the Republican Party, but congressional opponents re-emerged in spring 1907, more vocal and agitated than before.

The rapid growth in the number and size of forest reserves during Roosevelt's administration further fanned the flames of western resentment. The president withdrew seven million acres for national forests in Colorado in the span of two years, which left one-sixth of the state under federal control. In addition to using the Forest Reserve Act, Roosevelt withdrew more land under the Antiquities Act in 1906, which gave the president the authority to protect areas of scientific or historical interest on federal land as national monuments. Roosevelt created four national monuments that year and eighteen in all. He withdrew from use sixty-six million acres of public lands that contained potential coal deposits for surveying and possible sale, and he created the first game preserves and wildlife refuges. Each new proclamation added to the perception that the federal government was locking up the land.

Critics also trained their sights on the most visible spokesman for these actions—Gifford Pinchot. He urged Roosevelt to protect more and more land, much of which would fall under his agency's control. Pinchot's enemies (and their number grew rapidly after 1905) labeled his policies and his handling of the reserves "Pinchotism." One Colorado newspaper, complaining of the federal government's "Russian policy" of taking away lands, dubbed Pinchot "Czar" and his rangers "Cossacks."[167]

Opposition came from even within the Roosevelt administration. Secretary of Interior Ethan Allen Hitchcock resented the Agriculture Department's encroachment upon his domain. Hitchcock also believed in following the letter of the law, whereas Pinchot and Roosevelt believed in taking action until Congress or the courts explicitly forbade doing so. Hitchcock disapproved of Pinchot's initiatives, such as leasing the public domain and permitting commercial ventures on public lands because Congress had not authorized them. When the secretary began accusing a powerful conservation supporter in the Senate of illegal actions on the public land in 1906, Roosevelt decided that Interior under Hitchcock had "utterly gone to pieces." The secretary was eased out of office early the next year in favor of James R. Garfield, a close friend of Roosevelt's and Pinchot's and a major supporter of the conservation movement.[168]

Early in 1907, Senator Charles W. Fulton, a leader of the Sagebrush Rebellion, attached to the Department of Agriculture's annual appropriations an amendment that would overturn the executive powers granted in the Forest Reserve Act of 1891. It stated that no forest reserve could thereafter be created or enlarged in Oregon, Idaho, Montana, Wyoming, Washington, or Colorado except by an act of Congress. Significantly, it did not place a ban on new reservations in California, Utah, and Nevada because of the strong support for watershed conservation in those states. Fulton did not seek to punish Roosevelt or the Forest Service, but rather to reassert Congress's equality with the executive branch. The Fulton amendment was part of a bill to rename the forest reserves as national forests and increase the Forest Service's budget by $1 million.[169]

Faced with no choice but to sign the appropriations bill, Roosevelt instructed Pinchot to take action. Already armed with the statistics and information on all remaining unreserved forested lands in the six states, Pinchot and his men drafted presidential proclamations that created or enlarged thirty-two national forests, adding sixteen million acres to the National Forest System. As each proclamation was completed, Pinchot later recalled, he personally walked it over to the White House for the president's signature and gave it to the State Department for secure keeping.[170] Only then did Roosevelt sign the appropriations bill and announce the creation of what were later called the Midnight Reserves. Of this incredible political coup, Roosevelt wrote, "The opponents of the Forest Service turned handsprings in their wrath; and dire were their threats against the Executive; but the threats could not be carried out, and were really only a tribute to the efficiency of our action."[171] Congress was indeed outraged, and the agency faced difficult fights over appropriations for the next few years.

Roosevelt sent Pinchot and other cabinet members to lay the matter of conservation before the people and assess the political fallout. Just as King Darius sent his faithful servant Daniel to the lions' den, Roosevelt sent Pinchot to the 1907 Public Lands Convention in Denver, called by Colorado state legislators to debate the constitutionality of the Midnight Reserves and held just two weeks after their creation.

After listening to an angry crowd rail against the administration and its concept of federal sovereignty for two days, Pinchot finally rose to speak. An explosion of jeers and catcalls greeted him, but he held firm. "If you fellows can stand me," he called out, "I can stand you," a retort

Old Mother Heyburn went to the cupboard,
To get her poor dog a bone,
When she got there, the cupboard was bare.
And so the poor dog had none.
To MB Pinchot — compliments of
W. C. Morris
Spokane Spokesman-Review

that quieted the hecklers. He then launched into his standard talk about how all stood to benefit by embracing the National Forest System because it ensured that there was "no question of favoritism or graft." Instead of resisting the government's policies, the convention needed to understand the critical relationship of the national forests to the nation's future and the beneficial effects of scientific management. When he finished, the crowd reportedly cheered "lustfully," but he had not swayed everyone. Many who followed Pinchot to the podium continued the harangue. In the end, however, the resolutions committee, inexplicably stacked with members who favored conservation, asked for a reevaluation of national forest legislation and regulatory activism, a far cry from the vitriolic public pronouncements.[172] Daniel had tamed the lions.

To win broader support for conservation, Roosevelt appointed the Inland Waterways Commission in March 1907 to report on water transportation, flood control, waterpower development, irrigation, and soil conservation.[173] The commission was the first of several proposed

Some westerners supported President Roosevelt's creation of the Midnight Reserves in 1907. One editorial cartoonist showed his support for Roosevelt and his discontent with Idaho Senator Weldon Heyburn, a close supporter of western timbermen and vehement opponent of the new reserves, by portraying Heyburn as Old Mother Hubbard who had no bone for her dog, Timber Syndicates. (USDA Forest Service)

by Pinchot to promote conservation. The Inland Waterways Commission is remembered chiefly because it presented the newly christened Conservation Movement to the nation. The commissioners invited the president to sail down the Mississippi River with them in October 1907 to assure press coverage. At the end of the four-day trip, Roosevelt called for a national conference of governors on conservation.[174]

Pinchot's Dismissal and the Immediate Aftermath

Congress, still seething over the Midnight Reserves, denied funding for the Governors' Conference on Conservation. Unfazed, Pinchot funded most of it himself, sparing no expense. He even paid to have rubber tips placed on the legs of the chairs in the East Room in order to reduce the noise.[175] The Governors' Conference generated the desired publicity and further educated the public about the benefits of conservation. Efforts to create a world conference on conservation ended when Roosevelt left office in March 1909. It would be nearly forty years before such a conference was held again— this time under the auspices of the United Nations.

The election of William Howard Taft, Roosevelt's handpicked successor, left Pinchot fearful for his

Close friends and avid outdoorsmen, Chief Gifford Pinchot regaled President Roosevelt with a hunting story as they sailed down the Mississippi River as part of the Inland Waterways Commission. The commission helped introduce the term conservation. (USDA Forest Service)

programs.[176] Taft lost no time in confirming Pinchot's suspicions. Wanting to placate the West by appointing one of their own, Taft replaced Interior Secretary Garfield with Richard Ballinger, the reform-minded mayor of Seattle and a former commissioner of the General Land Office. When Ballinger was a member of the Public Lands Commission, he and Pinchot had crossed swords over his support of private development on public lands.[177] After taking charge of the Interior Department, Ballinger sought to curb the activities of the Reclamation Service and restore the majority of the waterpower sites withdrawn by Roosevelt at the end of his administration without waiting for the full evaluation of the sites. Almost as shocking to Pinchot was Ballinger's demand that Pinchot follow protocol to work with Interior personnel. This meant that Pinchot had to go through his boss, who went to Ballinger, who went to his agency heads.

Pinchot appealed to Taft to modify the secretary's decision about the waterpower sites. The resultant meeting between Taft and Ballinger made a splash in the national papers, with predictions

"I'll tell Teddy on you!"

BALTIMORE SUN

of Ballinger's dismissal.[178] That failed to materialize, but the press soon had a livelier story concerning Alaska coal fields located partly on the Chugach National Forest. At issue was whether Ballinger had tried to help a group he once represented obtain title to them. The story took on an almost sinister appeal when the press learned the claimants had an illegal agreement with the Morgan-Guggenheim business syndicate, a notorious trust believed to have a stranglehold on the Alaska economy. What erupted that summer and continued for the next eighteen months became known as the Ballinger-Pinchot controversy.[179]

Always the opportunist, Pinchot went public with his criticism of Ballinger's handling of the applications for the coal fields. When Taft imposed an order of silence on all government officials, Pinchot circumvented it by having Senator Jonathan P. Dolliver read a letter in the Senate floor detailing the issue and criticizing the president. Pinchot was willing to play the martyr to serve

As relations between Chief Pinchot and Secretary Ballinger deteriorated in 1909, it became apparent to many in the press that relations between Pinchot and President Taft were also crumbling. Taft fired Pinchot a short time later for insubordination. (USDA Forest Service)

the cause of conservation, and sure enough, Taft promptly fired him and Associate Chief Overton Price for insubordination on January 6, 1910. Philip Wells, the agency's law officer, resigned in protest soon thereafter.

Pinchot viewed the ensuing congressional investigation as nothing short of a defense of "the future of Conservation and the Forest Service. That was the heart of the matter."[180] He made the most of the proceedings by playing to the press in the gallery. Although the Senate exonerated Ballinger and Taft gave him his full support, with his reputation tarnished, Ballinger resigned in March 1911. Taft returned the coal claims to the public domain, but the political damage to his administration had already been done.

Pinchot reveled in his dismissal. It freed him to criticize the Taft administration as a private citizen and to promote conservation as he saw fit. A year earlier he had formed the National Conservation Association, which relied heavily upon his personal fortune for operating funds, to lobby Congress and educate the public about conservation. After his dismissal, NCA provided Pinchot a convenient platform to speak out on forestry issues in particular. Not surprisingly, a major focus of the association's criticism was Taft, who was already reeling from a fight over tariff reform and an insurgency within his own party; the Pinchot-Ballinger controversy further staggered Taft's presidency.[181]

Pinchot's dismissal shattered Forest Service morale. Most of the men had taken up forestry after reading his articles or working as student assistants. Many had been to his house in Washington for Society of American Foresters meetings or to Grey Towers while attending Yale; they recalled no greater thrill than when the chief came through on an inspection trip and spent time getting to know them and their professional problems. Over the years, they had come to share his vision for the country. Pinchot's dismissal put many in the Washington office in a fighting mood, willing to resign as a show of loyalty to their deposed boss. Pinchot visited the office the day after his firing to urge the men to be loyal to the Forest Service, not to him.

Rumors immediately surfaced that Pinchot would be replaced by a nonforester, A. P. Davis of the Bureau of Reclamation. Pinchot and Herbert "Dol" Smith, the agency's speechwriter and Pinchot's close adviser, conspired to get the best forestry man available, Henry S. Graves, the former associate chief and now dean of the Yale Forest School, as chief. They passed over Pinchot loyalist Albert Potter, who was serving as acting chief, because he, too, was not a forester. Instead, they maneuvered Graves's name in front of Taft and persuaded Graves to accept the job if offered. Reluctant to leave Yale, he took a one-year leave of absence beginning on February 1, 1910. At the end of the year, a satisfied Taft asked that his leave be extended. Graves decided to resign outright from Yale and continue his work until the "Forest Service and the Service policies are safe."[182]

A quiet, intense, plain man, Graves was temperamentally better suited for the academic life or solitary fieldwork than the political maelstrom of the Washington office. As associate chief,

he had proved a capable administrator and disciplinarian and a dependable second-in-command, which allowed Pinchot to lobby Congress for more appropriations and the transfer of the forest reserves.[183] Graves had been more comfortable in the role of dutiful subordinate.

An apt description of the contrast between the two men came from a western hotel clerk. When forester Thomas Woolsey tried to find Graves and Pinchot at their hotel, the clerk told him, "There was a tall man (Pinchot) and a short man that walked fast and sweats all the time."[184] Unwittingly, the clerk had made an astute observation: Pinchot strode through the halls of power among the nation's political and cultural elite and reveled in the job; Graves, from a middle-class background, was never comfortable as chief and preferred teaching.

Pinchot's firing jeopardized the future of the Forest Service. The first thing Graves had to do was make amends with Secretary of Agriculture Wilson. Pinchot had contravened Wilson's orders by criticizing Ballinger and had spent, in the secretary's opinion, too much time on nonforestry matters. Wilson wanted to restore discipline and ensure that the Forest Service focused on managing the land while avoiding further controversy. He required forest officers to obtain his permission before making speeches, for example. Wilson also launched a private investigation of the agency, bypassing Graves to correspond directly with field personnel. When Graves found out, he appealed to Taft and spoke directly with Wilson, offering loyalty in return for trust. Wilson agreed, and the relationship immediately improved and remained satisfactory. [185]

Chief Henry Graves (second from left) and Secretary of Agriculture James Wilson (middle) visited the Harney (now Black Hills) National Forest in South Dakota, 1912. Graves had to work hard to gain Wilson's trust when he took over the Forest Service after Chief Pinchot's dismissal in 1910. (USDA Forest Service)

The Forest Service had suffered from benign neglect while Pinchot had been off serving on various government and conservation commissions, generating publicity for the cause and resentment on Capitol Hill. With the flamboyant and independent-minded Pinchot out of the way, several western congressmen, at the behest of waterpower and mining interests, sought to transfer the national forests to the Interior Department or to the states. Such a transfer would jeopardize the headwaters of navigable streams, but—worse for Graves—place land management back in the hands of political appointees. To stop the effort, Graves decided the best course was to demonstrate that his agency was the best qualified for the job. He succeeded, but expended vital administrative energy fighting numerous transfer bills.

Congress left the forests under Forest Service control but continued to attack its management and its budget. Pinchot had promised Congress in 1905 that he would "make forestry pay" within a few years. Now Congress demanded to know how the agency spent its money and why expenditures exceeded income year after year. Graves and Associate Chief Albert Potter welcomed the congressional inquiry because it proved the Forest Service had managed its funds with great care. The decentralization initiated by Pinchot played a large part in cost cutting— travel expenses dropped from twenty-two percent of the appropriation in 1900 to less than seven percent in 1910—but a culture of frugality born of low appropriations better explains why the report reflected so well on the agency.[186] The agency learned to make do with even less when in 1911 Congress slashed the agency's budget for "permanent improvements" by more than fifty percent. With less money for roads, trails, telephone lines, and other infrastructure for forest management and firefighting, agency leaders demonstrated they were capable of good business management—and would henceforth respond to congressional wishes in order to protect the agency's budget.

Scientific Research

Four months after taking charge of the Forest Service, Henry Graves attended the official opening of the Forest Products Laboratory in Madison, Wisconsin, on June 4, 1910. For Graves, the opening was a rare bright spot in a year of tumult and challenge.[187] The Forest Products Laboratory quickly became central to the conservation movement because the better utilization of wood meant less waste, and more products from fewer trees.

A major objective for the lab was to develop methods to reduce logging waste and improve lumber production methods in sawmills. Researchers contributed to a basic understanding of wood chemistry and structure and improved wood preservatives and manufacturing processes, such as making plywood less expensive. During World War I, they tested the strengths of spruce for propellers, wings, and fuselages, and designed crates for shipping war supplies overseas. Curing and kiln technology for wood seasoning improved dramatically because of focused research as well.[188]

The decision to build the laboratory provided another example of the agency's willingness to meet shifting priorities. When Pinchot took office in 1898, he initially downplayed the emphasis his predecessor Bernhard Fernow had placed on research. Pinchot avoided the term research but continued some of the work as "investigations," in part because the latter connoted action. To court congressional support to transfer the reserves, he constantly emphasized the practicality of the division's work and the economic benefits of forestry. In 1901, the same year the division achieved bureau status, Pinchot revived forest products research under the Office of Special Investigations.

The Bureau of Forestry cooperated with other bureaus conducting research and contracted work to forestry schools, but it also contracted with prominent scientists. The agency funded forty-two individual investigations, including chemist Charles H. Herty's work on naval stores and plant pathologist Herman von Schrenk's study of the conditions, causes, and prevention of decay in timbers. In 1908, McGarvey Cline, chief of the Office of Wood Utilization, suggested forging a cooperative agreement with a university as the best way to establish a centralized testing facility for all wood products scientists. After a year of searching for a site, Pinchot finally settled on the University of Wisconsin at Madison as the site of the $30,000, 13,000-square-foot Forest Products Laboratory.[189]

Several years of advocacy by Raphael Zon, the head of the Office of Silvics, contributed to Pinchot's support for the lab. Zon's early experiences underlay his concern for the social goals

In 1932, the Forest Products Laboratory moved into its third and much larger facility not far from the original building on the University of Wisconsin campus. The first and third buildings are now in the National Register of Historic Places. (USDA Forest Service – Forest History Society)

Following the devastating fires of 1910, Raphael Zon, the director of the Forest Service's Office of Silvics, established the Priest River Experimental Station in the Idaho panhandle. Zon (seated, in white shirt) personally selected the site and helped build the facilities, including housing, a laboratory, and a seedling nursery. Construction materials, research equipment, and all other supplies had to be brought in by wagon. Priest River is one of seventy-seven experimental forests and ranges. (USDA Forest Service – Forest History Society)

of forestry, an approach that influenced the thinking of those around him. A socialist who had escaped from Czarist Russia to avoid prison, he had studied forestry under Bernhard Fernow at Cornell before joining the Bureau of Forestry, where he pioneered the study of the relation of forests, stream flow, and flood control. Although not a Yale graduate like most of Pinchot's inner circle, Zon nonetheless became a lifelong friend and ally of Pinchot. He headed the Office of Silvics from 1907 to 1914 and the Office of Forest Investigations from 1914 to 1923.

Shortly after being hired in 1901, Zon began the "objective gathering of accurate facts upon which to base sound programs of management of the nation's forest resources." Pinchot called Zon's small office "the first cradle and treasure house of forest research in America."[190] When Zon took over the Office of Silvics, he pushed the Forest Service to implement high-quality research on all phases of forestry. Zon's forceful calls for separating research work from administration and acerbic relations with his superiors, however, got him in trouble on occasion and contributed to his transfer in 1923 to the Lake States Forest Experiment Station, which he headed until 1944. He made use of his position as editor of the *Journal of Forestry* to question Forest Service policies. In the 1920s, he joined Pinchot in advocating for Forest Service regulation of private timberlands, which put him at odds with agency leaders.[191]

In May 1908, Zon pointed out to Pinchot the benefits of German research conducted on experiment stations since their creation in the 1870s. Pinchot agreed with Zon's suggestion about the need for research stations, and in August 1908, the Forest Service established the first one in Fort Valley, Arizona, on the Coconino National Forest. Zon selected the site for the station—he even helped build it—and then turned it over to its first director, Gus Pearson. It was the first of many stations placed in each silvicultural region under Zon's supervision. Within five years of opening Fort Valley, experiment stations were up and running in Idaho, Minnesota, Colorado, Washington, and Utah, along with dendrological and seed testing laboratories in Washington, D.C., two regional products labs in Wisconsin and Washington State, and the Forest Products Laboratory in Madison.[192]

One of the primary purposes of the national forests established by the Organic Act was to "maintain favorable conditions of water flows." The Forest Service maintained that forests were essential to regulating floods, and the Army Corps of Engineers argued that levies and dams were the only way to control water, but the subject had never been studied in the United States. The Wagon Wheel Gap Experiment Station on Colorado's Rio Grande National Forest was set up in 1911 to study stream flow and erosion, and the research conducted there proved less conclusive than the Forest Service had hoped.

More importantly, the results proved Zon's fear and revealed the need to make research independent of administration. Often, administrators in the Washington office and the district foresters' offices wanted research to support policy. Researchers operated with an implicit threat that they would lose funding if research challenged policy. At the district level, scientific investigations were subordinate to administrative whims, where "local problems of immediate importance" sometimes disrupted research efforts. Impatient administrators wanted researchers to produce results quickly. Districts even transferred incompetent administrators to experiment stations, where their inadequacies as researchers, too, discredited the research program.[193]

By 1915, there were many research efforts but little coordination between them, and as Zon had feared, credibility of the program was suffering. As a remedy, Graves separated research from administration by establishing the Branch of Research in 1915. Creation of Research completed the basic structure of the modern-day Forest Service, giving the agency its third main division. In addition to Research, the agency had the National Forest Administration to handle national forests and (later) national grasslands, and the Branch of State and Private Forestry, which Pinchot created in 1908 to support cooperation on lands outside the national forests.

Graves passed over Raphael Zon for chief of Research and instead appointed Earle H. Clapp, who held the position until his appointment as associate chief of the Forest Service in 1935. Full independence for the branch came slowly, and it took Clapp thirteen years to achieve it. The branch struggled toward autonomy because some research efforts remained decentralized long after its establishment. Range investigations remained under the grazing administration because of former stockman Albert Potter's insistence; dendrologists reported to the chief through their supervisor, George Sudworth; and all of Research reported to Silviculture and not to the chief. Furthermore, the success of the Forest Products Laboratory diverted funds from field research.[194] After more than a decade of lobbying by Clapp to improve the situation, Congress passed the McSweeney-McNary Act of 1928. The act legitimized the experiment stations, authorized broad-scale forest research, and authorized an appropriate level of funds to ensure success. Research now stood on a par with other forestry activities and was truly independent of administrative whims.[195]

In the 1930s and 1940s, funding for Research declined but the number of facilities expanded. New Deal programs like the Civilian Conservation Corps and Works Progress Administration provided labor and materials to construct new facilities. By 1935, the Forest Service had forty-eight experimental forests and ranges. Funding increases came during and after World War II, largely from defense contracts, and large numbers of professional scientists entered the Forest Service. Dozens of research centers—separate from the forest experiment stations—opened in the years immediately following the war. The structure moved senior researchers out of the labs and into administration, leaving the less experienced and lowest paid employees to do the research.

Research Chief Vernon L. Harper, appointed in 1951, recognized the shortcomings of the Forest Service's research program. The rapid increase of research efforts after the war resulted in some superficial research, leaders with too much autonomy, and administrative redundancy in some experiment stations. The Forest Service leaders, at Harper's suggestion, opted to streamline responsibilities. In the late 1950s, Harper adopted a project-based organization that allowed transfer of many administrative tasks to the station and increased the depth of research.[196]

In the late 1950s and early 1960s, increased demands on the nation's forests and growing public concern about the quality of forest management led the Forest Service to rethink its position on cooperative agreements with universities. The agency had phased out laboratories located at universities years before. When the Multiple Use–Sustained Yield Act of 1960 further expanded research needs into outdoor recreation and wildlife management, the Forest Service got the McIntire-Stennis Act passed in 1962 to foster financial cooperation with forestry schools located at the nation's land-grant universities.[197] Its passage also assured the Forest Service of a steady supply of scientists with graduate degrees. Much of this field research informed forest management decisions. Mandates of the Endangered Species Act of 1973, the Clean Air Act of 1970, the Clean Water Act of 1972, and the National Forest Management Act of 1976 added to the need for research.

The Forest and Rangeland Renewable Resources Research Act of 1978, which supplanted the McSweeney-McNary Act of 1928, authorized renewable resources research on national forests and rangelands, including research relating to fish and wildlife and their habitats. With the emphasis on ecosystem-based management and collaborative stewardship in the late 1980s and early 1990s, research expanded its efforts yet again. At the same time, however, budget cuts forced a reduction in personnel (from 964 scientists in 1979 to 720 in 1989) and forced consolidation or closing of projects and laboratories. The number of research units dropped from 247 to 190. Though the authorization for the McIntire-Stennis research program is 50 percent of the Forest Service budget, only about 13 percent was appropriated in the 1980s. Increasingly, the agency turned to cooperative research with universities to make up some of the difference. Since then, funding has continued to fall short of budgetary needs in many areas. Yet Research does what it always has—making enormous contributions as science knowledge and technology demands broaden and grow despite budget limitations.[198]

The Year of the Fires

The opening of the Forest Products Laboratory and Gifford Pinchot's dismissal would have been enough to make 1910 a memorable year in Forest Service history. However, 1910 was the "Year of the Fires." When the fire season of 1910 ended, more than five million acres had burned, as many as eighty-five firefighters were dead, and a new folk hero, Ed Pulaski, had emerged. Moreover, the catastrophic fires of 1910 had a lasting impact on federal fire policy.

In 1898, Pinchot had his staff study the history of forest fires to better understand damage. They catalogued more than five thousand fires since 1754 and estimated that fire cost the nation $20 million a year. The fires included those that struck Maine and New Brunswick in 1825 (approximately 3 million acres), Wisconsin in 1871 (1.28 million acres and more than 1,400 lives lost at Peshtigo), Michigan in 1881 (1 million acres and 169 lives), and Wisconsin and Minnesota in 1894 (several million acres and 418 lives in Hinckley, Minnesota). After the investigation, fires that stretched from upstate New York to Maine in both 1903 and 1908, and the Yacolt fire of 1902 (a series of 110 or more larger fires spread over 1 million acres in western Oregon and Washington) contributed to the development of state fire organizations and the establishment of private timber protective associations through the Weeks Act of 1911. They also provided vivid evidence for conservationists of the need for fire prevention and galvanized the Forest Service.

Pinchot, Graves, and many other foresters believed the use of fire for clearing land was "enormously destructive" and should be eliminated to protect timber resources.[199] The burning of undergrowth in the South was likewise widely condemned. European management models offered little guidance because foresters there had largely excluded fire from their forests. The Forest Service's position rested mostly on the belief that the public could not understand the difference between good and bad fires, so the best policy was not allowing fires at all. Grossly understaffed and lacking the technology and resources to fight fires over the vast national forest holdings, the Forest Service maintained that education was the best tool, preaching fire control and suppression.[200] Those who questioned why received a harsh reminder in 1910.

After an unusually dry winter and spring throughout much of the country, the fire season started early when the first flames broke out in April on the Coeur d'Alene National Forest in

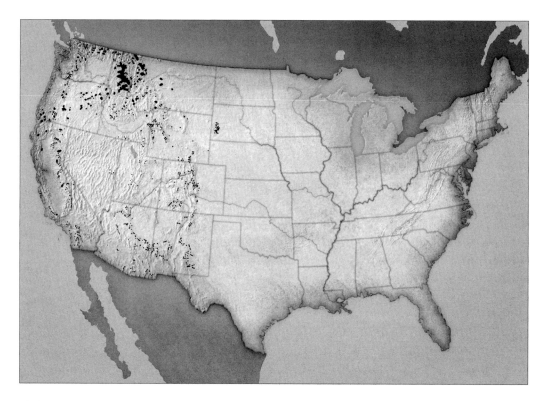

The fires of 1910 in the national forests of the western United States. The map does not reflect fires on state, private or other federal lands, which could double or triple the total area. (Fred Plummer, *Forest Fires*, Forest Service Bulletin 117, 1912)

Idaho, in District (later Region) 1. The Lake States eventually lost more than a million acres of forest to fire that year. The federal government dispatched thousands of troops to help battle the numerous western fires. On August 20 and 21 came "the Big Blowup." Several individual fires in and around the Idaho panhandle and western Montana converged and exploded with such ferocity that rangers recalled the fire as a hurricane, a vast conflagration, an awe-inspiring scene, a roaring furnace, or a threatening hell. Smoke stretched as far as New England, and soot traveled all the way to Greenland.

Of the many stories and heroes from the Big Blowup, one ranger's story stood out among them. Ranger Ed Pulaski, who later developed the eponymous firefighting tool that combines an ax with a mattock, gained fame for his courage and quick thinking. Although he lost six men, his simple desire to get home to his wife and daughter struck a universal cord with the public and launched him into Forest Service myth almost before the fires had stopped burning. "On August 20 a terrific hurricane broke over the mountains," Pulaski wrote twelve years later,

> *it picked up the fires and carried them for miles. The wind was so strong that it almost lifted men out of their saddles, and the canyons seemed to act as chimneys, through which the wind and fires swept with the roar of a thousand freight trains. The smoke and heat became so intense that it was difficult to breathe....*
>
> *Under such conditions, it would have been worse than foolhardy to attempt to fight the fires. It was a case of saving our lives. I got on my horse and went where I could, gathering men....*

I finally collected forty-five men. My voice was almost gone from trying to call above the noise of the fire and wind, but I finally succeeded in making them understand that if they would seize blankets from the camp stocks and do just as I told them, there was a chance of our saving our lives; otherwise they would be burned to death….

My one hope was to reach an old mine tunnel which I knew to be not far from us. We raced for it. On the way one man was killed by a falling tree. We reached the mine just in time, for we were hardly in when the fire swept over our trail. I ordered the men to lie face down upon the ground of the tunnel and not dare to sit up unless they wanted to suffocate, for the tunnel was filling with fire gas and smoke. One man tried to make a rush outside, which would have meant certain death. I drew my revolver and said, "The first man who tries to leave this tunnel I will shoot."

I did not have to use my gun…

The men were in a panic of fear, some crying, some praying. Many of them soon became unconscious from the terrible heat, smoke, and fire gas. The wet blankets actually caught fire and I had to replace them with others soaked in water. But I, too, finally sank down unconscious. I do not know how long I was in this condition, but it must have been for hours. I remember hearing a man say, "Come outside, boys, the boss is dead." I replied, "Like hell he is." I raised myself up and felt fresh air circulating through the mine. The men were becoming conscious. It was 5 o'clock in the morning…

We counted our number. Five were missing. Some of the men went back and tried to awaken them, but they were dead. As the air outside became clearer, we gained strength, and finally were able to stagger to our feet and start toward Wallace [Idaho]. We had to make our way over burning logs and through smoking debris. When walking failed us, we crawled on hands and knees….

We finally reached Wallace and were put in the different hospitals. Those who died were later brought out on pack-horses. Part of Wallace had burned in that same fire, so when my injuries were dressed I insisted upon going to my home, to make sure that my wife and little daughter were all right. I got a man to lead me, for the world was black to my eyes; but when I found my home and

Ed Pulaski, whose quick thinking saved the lives of most of his men, suffered medical problems caused by the fire for the rest of his life. Pulaski never received compensation for his expenses because of bureaucratic mistakes. His essay about his 1910 experience won an *American Forests* contest, and he used the $500 prize to help pay medical costs. (USDA Forest Service)

Wallace, Idaho, after the fires of 1910. The eastern third of the town burned after fire spread to a gasoline storage tank. Wallace was one of several western towns severely damaged in the fire. (USDA Forest Service)

family safe, they sent be back to the hospital, where I stayed for nearly two months with blindness and pneumonia. My experience left me with poor eyes, weak lungs, and throat; but, thank God, I am not now blind.[201]

In September, with fires still smoldering out West, the Forest Service administration began mop-up operations in the forest and on its reputation. In numerous articles and reports issued over the next few years, foresters contended that the widespread destruction was not their fault, and that a lack of funding and manpower had left them ill prepared for the battle. Instead of studying the impact of fire on the landscape, the agency went on the offensive to establish a policy of fire suppression. To aid the cause, the magazine *American Forests*, published by the American Forestry Association, published a special issue in which several foresters described what had happened in August around the country and what lessons they had drawn. Most took the opportunity to say they could have done more if properly outfitted.

In a brief introduction to the special issue, Chief Graves labeled fire as "a great obstacle in the way of the practice of forestry," whether public or private. He declared fire prevention the best policy to protect America's forests and urged state and private cooperation so that all forestland had protection.[202]

Ed Pulaksi and his men took refuge in an old mining tunnel, photographed shortly after the fire. (USDA Forest Service)

Ferdinand Augustus "Gus" Silcox, the quartermaster general under District Forester William B. Greeley at the time of the Big Blowup, was one of the contributors to the special issue on fire. Like many of his agency colleagues, Silcox had read about forestry in an article in the *Saturday Evening Post*, signed up as a student assistant before going to the Yale Forest School, and then entered the Division of Forestry a "little G.P.," as many early foresters called themselves in honor of their boss. Silcox's exceptional intellectual gifts and ability to resolve disputes in face-to-face meetings helped him move up quickly through the ranks to become associate district forester in Missoula in 1908. Two years later, at a critical moment in the agency's history, he was head of the Forest Service's quartermaster corps.[203] Silcox had the difficult task of equipping, feeding, and paying the men—some ten thousand by mid-August—on the frontlines, a largely thankless job that convinced him that he could have saved more forest if only he had had more equipment. His experience in the 1910 fires left him obsessed with eliminating fire to prevent a recurrence of the disaster.

Silcox's article described the fire and the damage, the herculean efforts to combat it, inadequacies of equipment, and the difficulties of transportation in the backcountry. More importantly, the article revealed how quickly the incident shaped Silcox's thinking about the place of fire in the landscape, and about how the Forest Service should handle it in the future.

For decades, pack trains were the most reliable way to transport supplies over difficult terrain. Today a part of agency lore, they are still used for backcountry camping trips in national forests.
(USDA Forest Service)

Like a good "little G.P.," he believed science, technology, and manpower were the answers, if only the Forest Service applied enough of each to the problem: "The inadequate trail systems…made it impossible to get to a great many of these fires" until they had become too large to be handled by a few men, but "with an adequate trail, look-out, and telephone system, and a sufficient equipment of tools, the fires can be controlled." He then rhetorically asked, "Is it worth while?" It was all a matter of economics. He estimated the timber lost in the national forests in northern Idaho and northwestern Montana at approximately eighty billion feet and worth $200 million, not counting lost wages to the community.[204]

Reactions to the fires were unexpected and diverse. Graves, Pinchot, and other conservation leaders viewed the fires as beneficial to the agency's cause. At the National Conservation Congress, held immediately after the Big Blowup, Graves said the conservation movement had succeeded in its public message. What remained for forestry was the "practical application" of its principles: protect the woods from fire, reduce waste, and ensure a future timber supply, in that order. The fires demonstrated what could happen if the Forest Service—and the nation—should shrink from that duty.[205]

Foresters like Elers Koch, who argued in favor of letting backcountry fires in Idaho and Montana burn themselves out, and Professor Herman H. Chapman of the Yale Forest School, who suggested that controlled light burns had a positive role in southern landscapes, found their opinions suppressed by the Society of American Foresters and the Forest Service. Not until 1943 would the Forest Service accept evidence that fire, if properly applied, benefited longleaf pines in the South. Even then, it refused to apply that knowledge to other landscapes. Until the 1970s, light burning remained "a political threat and not…a management technique."[206]

To Progressives like Pinchot, fire was also a socioeconomic threat. It destroyed merchantable timber and threatened a resource vital to the nation's economy and even social fabric. The 1910 fires reinforced this belief, and fighting all fires, regardless of location or season, became one of the top priorities of the Forest Service. When talk of dismantling the agency arose shortly after he took over, firefighting gave Graves the best argument for keeping national forests under federal control, not to mention a reason for the continued existence of the Forest Service.

Congress responded to the request for more money to fight fires by approving the Weeks Act in 1911. Section 2 of the law authorized federal matching funds (initially $200,000) for states with forest protection agencies that met government standards. The law fostered the creation of new state programs and the improvement of existing ones, and allowed fire commissions to use the money on state and private lands, just as Graves had called for in his *American Forests* article. By the time Graves left office in 1920, he listed twenty-three states as participants in the program, which meant they also shared the burden of fire control, even on private lands. By then, nearly 95 million acres of national and state forests and about 160 million acres of private forests had fire protection because of the Weeks Act.[207] Within the Forest Service, the act also increased the standing of State and Private Forestry.

The Weeks Act also provided money for the purchase of private lands for additional national forests as part of the effort to protect watersheds. By passing it, Congress for the first time engaged in purchasing private land for public use. After some initial difficulties, the National Forest Reservation Commission, created to oversee the process, had purchased nearly twenty million acres by 1961, almost entirely in the East.

Establishing Fire Policy

William Greeley, who succeeded Graves as chief in 1920, had been the District 1 forester at the time of the 1910 fires. Like his assistant Gus Silcox, he, too, wanted to eliminate fire from the landscape. Firefighting, Greeley said, was "a matter of scientific management, just as much as silviculture or range improvement." It was the bedrock of forestry and conservation.[208] He dedicated the Priest River Forest Experiment Station to fire control research, and put Harry Gisborne in charge of fire research.

The "Gisborne Era" of fire research lasted from 1922 to 1949, when Gisborne died of a heart attack while investigating the destructive Mann Gulch Fire in Montana, which trapped and killed thirteen smokejumpers. Fire research during the 1920s was subordinate to administrative needs and focused on fire control rather than fire itself. Gisborne's personal creed—that research was a waste of time unless it addressed real problems and could produce results for immediate application—guided the work at Priest River. Research focused on three principle interests: measuring and forecasting fire conditions, predicting lightning-caused fires, and long-range statistical forecasting of rainfall. This led to the

State and Private Forestry

As chief of the Division of Forestry, Bernhard Fernow spent considerable time answering queries from homeowners and forest landowners about their trees and shrubs. Since then, the Forest Service and its predecessors have been providing cooperative assistance to landowners. In 1908, Pinchot recognized the Forest Service's obligation to the private sector when he formally established the Branch of State and Private Forestry, now one of three branches (Research and the National Forest System are the other two).

Chief Henry Graves envisioned cooperation in three forms: Advising states in establishing forest policies, assisting them in surveying their forest resources (primarily timber), and helping forest owners with practical forestry problems. Section 2 of the Weeks Act of 1911 codified these responsibilities. Under the Weeks Act, the Forest Service could combat fire even if located on private or state land. The Clarke-McNary Act of 1924 greatly expanded cooperation with private and state interests, especially on taxes and fire. The Cooperative Forest Management Act of 1950 extended management assistance to all classes of forest ownership. Since the Cooperative Forestry Assistance Act of 1978, the Forest Service's cooperative efforts have involved working with their state and private counterparts in forest pest and disease control, urban forestry, and non-industrial private forest landowners.

Source: Gerald W. Williams, The USDA Forest Service—The First Century *(2000), 27–29.*

S-52

FIRE DANGER

⟨3⟩

RELATIVE   WIND   DAYS SINCE
HUMIDITY VELOCITY   RAIN

⟨38⟩   ⟨3⟩   ⟨0⟩

FUEL   PREDICTION   WIND
MOISTURE TO-MORROW DIRECTION

⟨9⟩   ⟨3⟩   ⟨NW⟩

FIRE DANGER
1. DORMANT   3. MODERATE DANGER
2. LOW DANGER 4. HIGH DANGER
5. EXTREME DANGER

Fire research pioneer Harry Gisborne developed in 1931 what he later called "my major research contribution" — the fire-danger meter. By integrating several factors such as fuel moisture percentage, wind velocity, and relative humidity, Gisborne created a system that measured probable fire danger easily understood by foresters and laymen alike.
(USDA Forest Service)

development of backpack radios, the fire-danger rating system, and other fire control tools and techniques. Because research focused on fire control, however, and not fire ecology, a better understanding of the role of fire in western forests would not emerge until later.[209]

As chief, Greeley also worked to improve the fire control provisions of the Weeks Act. Congress obliged Greeley with the Clarke-McNary Act in 1924, which further expanded the ability of Congress to purchase lands, and encouraged reforestation efforts and closer federal, state, and private cooperation for fire control. The McSweeney-McNary Act of 1928 fostered cooperation and, more importantly, supported a $3.6 million research program that included funding for fire research. The new research monies gave the Forest Service a virtual monopoly over federal research on fire. It also unintentionally further suppressed critics such as Herman Chapman and Elers Koch.[210]

In 1933, Gus Silcox was appointed Forest Service chief, which gave him the opportunity to combat fire on a national scale.[211] As district ranger, Silcox implemented fire control reforms before the dry fire season of 1914. His experience convinced him the agency could fight fires, if only properly outfitted to do so. Devastating fires in the Selway Mountains of Idaho in 1934—the worst since 1910—reinforced in Silcox's mind the need for preventive action and triggered renewed debate about fire policy. A review board convened by the Forest Service in September 1934 outlined three possible strategies: aggressively suppress fire to "keep every acre green"; continue trail and road development to enhance firefighting; or withdraw from aggressive fire control in the backcountry. The board dismissed the third option because public opinion, shaped by two decades of Forest Service propaganda, was opposed to a let-burn policy. An additional report stated unequivocally that shortcomings in fire control resulted from not doing enough. Silcox embraced the first option.[212]

Silcox put the Civilian Conservation Corps, a New Deal public works program, to work thinning forests and building roads and trails in what he hoped would be the final assault on fire. In 1910, he had said he could virtually eliminate destructive forest fires if given properly supplied manpower.

Under his leadership as chief, his plan seemed to work. The amount of land burned dropped steadily in the 1930s, even as the federal government added more and more land to the National Forest System during the New Deal.

Silcox also instituted the "10 a.m. policy," which stipulated that every fire should be controlled by ten the morning following its report. If the fire was not under control, then fire officers had until ten the next morning to control it, and so on, for weeks if necessary, until they extinguished the fire. Regardless of origin or location, Forest Service personnel treated fires everywhere with equal force. An increase in equipment, such as lookout towers, telephone lines, spotter aircraft, and by the end of the decade, smokejumpers, aided their efforts.

Returning Fire to the Landscape

Federal land managers enforced the 10 a.m. policy until the National Park Service, wanting to manage parks as ecosystems in part by restoring fire to the landscape, refused to implement it in 1968. The Forest Service modified policies over the next ten years before it adopted a program similar to that of the Park Service. Beginning in 1978, natural fires would be allowed

In the early years, fire watch-towers varied in comfort and ease of accessibility, as shown on the following pages.

Shown above is Twin Sisters Fire Lookout Tower, Colorado National Forest, 1917.
(USDA Forest Service)

Fire lookout tree on Turkey
Knob, Webster County,
West Virginia, 1919.
(USDA Forest Service)

to burn in wilderness sites, and controlled burning would be permitted to reduce fuels and improve habitat.

Reintroducing fire was not a simple matter of letting fires burn, or of setting controlled burns. Forests that had not been allowed to burn had become diseased and were dying, susceptible to catastrophic fires. The Yellowstone Fires of 1988, followed by fires in 1994 that killed thirty-four firefighters, burned two million acres of forest, and consumed $965 million in emergency fire funds, exposed the fire control policy for what it had been—misguided at best.[213] Certainly, after seventy-five years, it was apparent that fire policy had been going in the wrong direction. Fires were becoming bigger and more dangerous because the fuel buildups burned with greater intensity.

In 1995, the relevant federal agencies, now including the Bureau of Land Management, agreed to suppress "bad" fires and allow "good" ones to burn. However, the urge to suppress

Smokey Bear

During World War II, concern about forest fires led to a forest fire prevention campaign in California that quickly became a nationwide cause. Between 1942 and 1944, fire prevention posters used wartime slogans and then the cartoon character Bambi before settling on a bear as its "spokesman." Designers were told to make the bear's "expression appealing, knowledgeable, [and] quizzical," and to include a hat that typified the outdoors and woods. Blue jeans were added later. They purportedly named the bear Smokey after Smokey Joe Martin, a heroic New York City fireman. Smokey first uttered his famous slogan, "Only YOU Can Prevent Forest Fires," in 1947.

In 1950, firefighters rescued a bear cub from the Capitan Gap forest fire in New Mexico. The badly burned bear was dubbed "Smokey" and became a living symbol of fire prevention. Smokey's popularity quickly grew. In 1952, a stuffed toy version of Smokey Bear came with a card that children could send in to become "Junior Forest Rangers." Within three years, 500,000 children had responded. By 1965, Smokey was receiving so much mail that the Post Office gave Smokey his own zip code. Until he died in 1975, the bear was the most popular attraction at the National Zoo in Washington, D.C. Since the 1970s, the fire-preventin' bear's message and its effectiveness have proved to be a mixed blessing for the Forest Service: the agency now faces opposition to its efforts to return fire to the landscape.

Source: Gerald W. Williams, The USDA Forest Service—The First Century *(2000), 84–86; and William Clifford Lawter, Jr., Smokey Bear 20252: A Biography (Alexandria, VA: Lindsay Smith Publishers, 1994).*

Smokey Bear has played a crucial role in the Forest Service's fire prevention campaign since the 1940s. He is the third most recognized figure in American popular culture, behind Santa Claus and Mickey Mouse. (USDA Forest Service)

had become ingrained in the Forest Service's culture, as well as that of American culture. Fifty years of hearing Forest Service icon Smokey Bear's warning, "Only you can prevent forest fires," had conditioned the general public to reject fire in any form—whether controlled, prescribed, or wild—on public land. Attempts to reintroduce fire through prescribed burns quickly proved as dangerous as fighting fire. "Controlled" burns that got out of control, such as one the National Park Service set near Los Alamos, New Mexico, which eventually burned nearly forty-seven thousand acres in May 2000, only reinforced the belief that fire might not belong in the landscape. At the very least, the question of how much fire to allow remains divisive.

The Forest Service and other land management agencies, and the nation as a whole, are at a divide—a firebreak, one might say. Excluding fire allows the continued accumulation of fuels. Where land managers have kept fire out, fire-intolerant species of trees and shrubs have entered the landscape. They can serve as ladders that enable flames to climb from the forest floor to the crowns of larger, older trees, leading to hotter, catastrophic fires that damage soils, cause erosion, and endanger human communities. Even though advanced technology and the rapid response of skilled firefighters combine to extinguish ninety-eight percent of wildfires in an initial attack, the total annual area of forest burned since the mid-1980s has been increasing.

Most fire experts and ecologists argue that reducing the level of hazardous fuels in the forests and rangelands (which also suffer from the exclusion of fire) is a necessity if society is to restore healthy landscapes and protect human communities. It is at this point that one finds the firebreak. As fire historian Stephen Pyne has pointed out, "We're not going to cut our way out of the problem and we're not going burn our way out. We can't suppress it and we can't walk away from it. What we need is to do lots of little things, mixtures of those kinds of practices, adjusted to particular sites."[214] Many in the timber industry call for aggressive logging to reduce fuel loads; the environmental community rejects that approach and says that the National Fire Plan of 2000 and the Healthy Forests Act of 2003 are ways to increase timber cutting with minimal oversight.

Caught in between is the public land manager. The manager has to walk a fine line between federal laws that call for the production of timber, forage, or minerals, such as the National Forest Management Act of 1976, the Taylor Grazing Act of 1934, or the Mining Law of 1872, and those that mandate the protection of wildlife habitat and water, such as the National Environmental Policy Act, the Endangered Species Act of 1973, and the Clean Water Act of 1972. Although the most thoughtful management plans may bring legitimate criticism, arguing factions and conflicting legislation frequently place the land manager in a no-win situation.[215] It is a position that the chiefs, from Henry Graves to Dale Bosworth, have known all too well when dealing with fire.

THE PRIMACY OF TIMBER
AND THE PROMISE OF SUSTAINED YIELD

A forester explains the merits of leaving small trees, October 1937. (U.S. Forest Service)

THE EVENTS OF 1910—THE DISMISSAL OF GIFFORD PINCHOT AND THE FIRE SEASON—LEFT THE FOREST SERVICE "FEELING CORRECT BUT BRUISED."[216] THE AGENCY, DOMINATED BY FORESTERS WHO HAD LITTLE TRAINING BEYOND TIMBER MANAGEMENT, MAINTAINED ITS MISSIONARY ZEAL TO PREVENT A TIMBER FAMINE. FORESTRY SCHOOLS—MANY OF THEM STAFFED BY FORMER STUDENTS OF HENRY GRAVES—AND THE SOCIETY OF AMERICAN FORESTERS INSTITUTIONALIZED FORESTERS' BELIEF THAT THEY, STEEPED IN TECHNICAL PRINCIPLES AND FREE OF INFLUENCE FROM POLITICS OR INDUSTRY, KNEW WHAT WAS BEST FOR THE LAND. AFTER THE FIRST GENERATION OF FORESTERS HAD PASSED FROM THE SCENE BY THE 1940S, THERE WAS LITTLE DISAGREEMENT OR DISSENT WITHIN THE AGENCY REGARDING TIMBER MANAGEMENT. THIS WAS A FOREST SERVICE, AND TIMBER WAS ITS TOP PRIORITY.

Other issues—wilderness protection and recreation, fending off the National Park Service's challenge to Forest Service hegemony over land management, the country's entry into World War I and the draining of manpower, expansion of the eastern forests under the Weeks Act, and passage of the McSweeney-McNary Act of 1928—arose but remained subordinate to timber. Because little cutting occurred on public lands before World War II, the timber policy debate initially focused on private lands. Chiefs Henry Graves and William Greeley debated with Gifford Pinchot and his followers about direct regulation of private timberlands to prevent timber famine and protect watersheds. The debate ebbed and flowed for more than forty years.

Crafting a Timber Policy

The Organic Act of 1897 authorized the secretary of Interior to establish regulations to promote "younger growth" on forest reserves, which were established in part to "furnish a continuous supply of timber." Logging was permitted but not required for forest management. Through a deal brokered by Gifford Pinchot when he was a forest agent working for Interior, the General Land Office made its first timber sale on the Black Hills Forest Reserve in 1898, to the Homestake Mining Company. The deal reflected Pinchot's desire to demonstrate how scientific forest management could meet industrial needs while conserving forests for future use.

By the time logging began under the contract, Pinchot was chief of Agriculture's Division of Forestry and provided Homestake with foresters to devise the working plan. The contract for fifteen million board feet to supply lumber for Homestake's mining operations required that no trees smaller than eight inches in diameter be removed, and that after the harvest the brush left behind be gathered in piles to prevent forest fires. The deal, wrote future chief of the Forest

Awarded to the Homestake Mining Company to supply timber for its mining operation, Timber Sale #1 provided Gifford Pinchot and the Division of Forestry its first opportunity to demonstrate the benefits of scientific forest management. The mining operation shut down in 2003.
(Black Hills National Forest)

Service Greeley, gave "the enthusiastic young foresters" in Agriculture's Division of Forestry the opportunity "to do real business in selective logging." Thus began the agency's timber program.[217]

Homestake set important precedents. Pinchot had persuaded a large timber user to stop its destructive logging practices and instead use scientific forest management practices for long-term benefits. The job also required a working plan, which after the transfer in 1905 was required on all national forests. The inclusion of an approximate timber yield was intended to prevent overcutting and to calculate and manage the rate of timber harvest consistent with yearly growth and prospective local needs. By the 1920s, detailed management plans for each timber-producing forest estimated the amount of timber that could be cut from "working circles," areas that contained enough timber and timber growth to support local forest industries.

Through the 1920s, however, the Forest Service made few timber sales, especially compared with the timber harvested on private lands. In fact, between 1905 and 1945, the annual national forest timber harvest averaged less than one billion board feet, represented only two percent of the nation's timber supply from domestic sources, and involved less than two percent of the total national forest area. The few sales the Forest Service did make in its first two decades of operation were usually to large companies like Homestake and were designed for working circles that covered entire drainages over several decades.[218]

With only a few large timber sales to oversee, the Forest Service from its inception in 1905 until around 1933 could afford to manage national forests in what it called custodial fashion. This involved protecting existing resources, improving depleted range and forage, and preparing for the day when private timber stocks had dwindled and public lands would supply the nation's timber. When that day came, the agency would "be in a position to impose conditions upon purchasers which tend toward good forestry," one forester wrote in 1910, "the enforcement of which today is impracticable."[219]

Until the expected timber famine arrived, the agency concerned itself with establishing a timber policy and changing private industry's indiscriminate logging practices on private land. Only a minority of foresters in the Forest Service believed they should be passive custodians. Most favored allowing some cutting on the national forests for two reasons. One was economics. Withholding public timber hurt local communities dependent upon it for jobs; cutting federal timber promoted community stability. Moreover, any harvest moratorium would deny the local counties the twenty-five percent of federal timber receipts due them by law. And when he persuaded Congress to transfer the reserves in 1905, Pinchot had promised that revenues from timber and grazing would make the agency self-sufficient in a few years. When that failed to materialize, he started fudging the numbers to show increasing revenues, as he had done at the Biltmore.[220]

The other reason involved the science of forestry. Cutting would place the national forests on a sustained-yield basis much sooner than if the agency waited until private timber could no longer meet the nation's needs. In short, foresters wanted to transform national forests from "wild" to cultivated forests. Old-growth (or "overmature") forests were unproductive because they grew slowly—a viewpoint that characterized federal forestry for much of the twentieth century—and would not last long once cutting began. Younger trees, however, increased the total volume of wood in the forest at a faster rate. By embracing sustained yield at the outset, the Forest Service would have a system in place to meet future needs.

By 1910, congressional critics wanted to know why timber sales did not generate more money, as Pinchot had promised. Timber sales from national forests then provided barely one percent of the total lumber output. To deal with the criticism and devise a timber policy, Chief Graves brought District Forester William Greeley in from Montana. Greeley had studied at Yale Forest School under Graves, who had pegged him early as a man of destiny in the Forest Service. "Greeley had the highest mark of any of our recent graduates," Dean Graves wrote to Overton Price in 1904 when Greeley applied for work with the Bureau of Forestry. "He is a special star and I recommend him for almost any work which may come along."[221] Greeley did not disappoint. He quickly rose to district forester by 1908 and was promoted to assistant forester in charge of the Office of Forest Management for his handling of the 1910 "Big Blowup" in the northern Rockies.[222]

Greeley got right to work. Timber, he noted, could not be permanently stored; it lost value rotting on the stump and wasted public property. Waste was the "very antithesis of conservation," he reported to Graves. To eliminate waste, Greeley proposed that the Forest Service increase timber sales but withhold sales when the market was low. Greeley believed approaching the

To support U.S. participation in World War I, the Spruce Division went to work harvesting timber in the Pacific Northwest (above). Meanwhile, the forest engineers trained in the nation's capital before being sent to France to log and cut lumber for fellow troops (facing page). (USDA Forest Service)

problem with industry's interests placed ahead of the Forest Service's silvicultural needs would have minimal impact on local private markets and could ultimately benefit lumbermen as well.[223]

Private land holders hesitated to cut government timber instead of their own because of the weak lumber market.[224] Lumber industry critics such as Gifford Pinchot blamed the weak lumber market conditions on industry, which still suffered from the boom-and-bust cycle that had afflicted it since the mid-1800s.[225] Pinchot also blamed industry's destructive logging practices for the impending timber famine, accused his former agency of being industry's dupe, and called for greater federal oversight of logging on private land.[226] The debate over federal control versus federal cooperation was joined.

Greeley's major Forest Service report, *Some Public and Economic Aspects of the Lumber Industry*, issued in 1917, refuted Pinchot's contentions. Greeley found that lumbermen bought huge tracts of low-cost timberland, believing that the prices would rise shortly and they would see a return on the investment. Unlike other real estate ventures, lumbermen had a limited time to hold on to the land, and many could not keep it long enough to turn a profit, which compelled them to sell in an already depressed market to pay off creditors. Glutted markets drove prices down further, creating a vicious cycle that lasted throughout the pre–World War I years. Meanwhile, the Forest Service struggled to justify its promotion of cutting on government lands in the depressed market.[227]

Greeley concurred with industry's view that overcapitalization and high interest rates hurt lumbermen and insisted that industry needed help, not criticism, from the Forest Service. His report affirmed the existing policy that was intended to benefit both industry and the public: avoid competing directly with lumbermen where the market was soft, withhold federal timber until private supplies were exhausted, solicit competitive bids for lumber, and protect national forests against fire and other disasters. Pinchot strongly disagreed with Graves's decision to maintain the status quo, saying it was a "whitewash of destructive lumbering."[228]

By the time Graves left office in 1920, the Forest Service had firmly cast its lot with timber management as the primary purpose of the national forests. Greeley curbed the modest timber

sale program set up by Graves in favor of long-term contracts that promised larger volumes of timber over longer periods. That gave lumbermen the incentive to invest in the capital required to bring stability to mill towns. It also left the agency increasingly dependent upon timber sales receipts for revenue, and the only way to ensure a steady flow of revenue was to increase timber sales. As historian David Clary has noted, "Most foresters concentrated on the methods needed to increase timber supplies, instead of on the multiple use of the forest. Achieving sustained timber harvests, accordingly, outweighed other responsibilities."[229] From the outset, economic considerations shaped management decisions.

Cooperation versus Regulation

The decision to increase cutting on national forest land, however, placed the Forest Service in a quandary. The agency needed to generate receipts and become self-supporting. Yet to do so often meant allowing lumber companies to use a technique the agency did not officially sanction—clearcutting—even though some of its own studies supported its use in certain circumstances. Early Forest Service studies had shown that Douglas-fir in the Northwest and lodgepole pine in the northern Rockies regenerated faster if the land was clearcut and all debris removed. Studies from the 1920s, however, returned mixed results, reflecting the difficulties of forest research and the lack of long-range data. The Forest Service favored selection cutting to stop further forest destruction, and clearcutting did not become the predominant method of timber harvest and regeneration until well after World War II. But in the Pacific Northwest, clearcutting Douglas-fir had become the norm before the war.[230]

In a decision strongly reminiscent of how it educated the public about fire, the agency was unwilling to differentiate between appropriate clearcutting and destructive clearcutting and

simply declared all clearcutting bad. Forest Service publications blamed private industry's use of clearcutting for existing forest health problems and rallied the public to its side. Talk of regulating private lands to control private industry grew louder.[231]

The Forest Service set aside its discussion about regulating private lands, however, when the United States entered the Great War in April 1917. The war left the agency scrambling to meet its responsibilities as scores of men left for military service. The U.S. Army formed two forestry regiments (the Tenth and Twentieth Engineers) and sent them to France to secure timber for American forces. Under the command of Graves and Greeley, who temporarily left the Forest Service to serve in the regiments, the engineers built and operated sawmills to cut timber for railroads and to line trenches for American soldiers.[232] At home, the U.S. Army Spruce Production Division deployed some thirty thousand troops in Washington and Oregon to harvest spruce for airplanes and Douglas-fir for ships. To fill personnel shortages and still maintain fire protection, the Forest Service hired women as clerks and lookouts. One woman even worked as a "patrolwoman" in the Pacific Northwest, covering her district on horseback and carrying camping gear for stays in the woods.[233]

When the war ended and operations returned to normal, the Forest Service resumed its warnings of a timber famine. In 1919, the agency drafted its national forest timber management plans based on that fear and inadvertently placed itself in a difficult position. Chief Graves called for an expanded state-federal cooperative program of fire protection that would also require the states to regulate logging on private lands in a Forest Service-approved manner; the agency would

To increase its effectiveness and range in patrolling national forests, the Forest Service used whatever best suited the terrain. Because most fires started along railways, a ranger could pedal a velocipede (left), or "speeder," or use a modified car on railroad tracks (facing page). (USDA Forest Service)

step in if the states did not enforce standards. Industry leaders balked. Much to Graves's irritation, Gifford Pinchot organized a Society of American Foresters committee that demanded federal regulation of all forestlands. Graves dismissed the proposal as containing "many socialistic features."[234] The following year, exhausted and ill, Graves left the Forest Service. Sixteen years after heralding Greeley as "a special star" for the agency, Graves recommended Greeley as his replacement.

Greeley did not share Pinchot's antipathy toward the timber industry or his conviction that unregulated harvesting posed the utmost danger to the nation's forests. When Greeley pushed a modified version of the Graves proposal, the two men exchanged barbs in the *Journal of Forestry* and before congressional committees. Pinchot suggested that the man who had created the Forest Service knew what was best for the nation and forestry, and he accused Greeley of cozying up to industry. Viewing the situation through the prism of conservation, he believed that the lumber industry clearcut the land to unload timber as quickly as possible, and he had no tolerance for their destructive practices. Through his allies on Capitol Hill, Pinchot managed to get several versions of a regulation bill to the Senate floor.[235]

But Greeley proved as adept as Pinchot at the political process. He drafted a bill and gave it to lumber lobbyists, who then brought it to supporters in the House. When both his and Pinchot's bills stalled, Greeley backed a successor to his own that had bicameral support, the Clarke-McNary Act. Greeley stacked the congressional hearings with supporting witnesses, and the bill sailed through with little opposition. By then, Pinchot, who was serving his first term as governor of Pennsylvania, could devote little energy to the fight.[236] When it came to working the levers

of political machinery, Greeley had learned from—and then defeated—the master. Greeley's cooperation with industry began a relationship between the Forest Service and industry that, despite some disagreements, was ultimately symbiotic.

A major step in the evolution of forest policy, the Clarke-McNary Act emphasized federal, state, and private cooperation by authorizing federal matching funds for state forest fire programs. States became equal partners with the federal government in running cooperative nurseries and related reforestation programs and providing technical advice and services to owners of small woodlots. Additionally, the new law expanded the kinds of land that could be purchased for national forests and permitted the Forest Service to accept gifts of land (often cut over) and exchange land with other federal agencies inside or near the forest boundaries to simplify management, a process known as "blocking up" the national forests. Overall, the Clarke-McNary Act gave the Forest Service many programs with which to wield influence over timber management both inside and outside the national forests.

When Greeley resigned in 1928 to become executive secretary of the West Coast Lumbermen's Association, Pinchot and his supporters saw it as proof that he had sold out to special interests years before. The move actually epitomized the interdependence of public and private forestry, and Greeley spent the rest of his career trying to promote conservation from "the other side."

Rangers used motorcycles on the Tahoe as early as 1915 for fire patrol. On the Tongass National Forest in Alaska, a boat served as a mobile ranger station and office. (USDA Forest Service)

Pinchot and other foresters resumed their criticism of private industry for clearcutting followed by inadequate reforestation shortly after Robert Y. Stuart was named chief in 1928. Stuart was another Yale Forest School graduate who worked the Big Blowup of 1910. He left the Forest Service to work as Pinchot's deputy on the Pennsylvania Forest Commission from 1920 to 1922 and replaced Pinchot when Pinchot resigned to run for governor. In 1923, Stuart took charge of Pennsylvania's new Department of Forests and Waters before returning to the Forest Service in 1927 as assistant forester in charge of public relations. His appointment as chief in 1928 gave Pinchot greater access to the Forest Service's leaders and reinvigorated the regulation debate, though Stuart was more willing to seek a compromise with industry.

Timber and the Great Depression

Lumber production reached a peak in 1926 and then began a downward spiral that lasted until the mobilization effort for World War II. Most of the timber cut for the construction and industrial boom of the Roaring Twenties had come from private holdings, but pressure to cut timber on national forests was increasing. Perennially shorthanded, the Forest Service found it could not properly supervise all sales, leaving some companies to police themselves on national forests. One company that had shared the agency's concern for future resources "nonetheless destroyed over half of the unmarked timber, two-thirds of the poles, and three-quarters of the seedlings in the harvest areas"; the figures were typical of most national forests.[237] The announcement in 1927 by one Virginia company that it would continue operations "with a view towards continuous production" by replacing clearcutting with selection cutting operations was the exception to the rule.[238]

Overcutting and the depressed timber market of the Great Depression together revived the issue of regulation. Critics charged that land-holding companies were liquidating assets to stay in business, leading them and the lumbermen to engage in destructive lumbering, waste, and premature cutting. The government had to contend with the loss of land from tax rolls and growing unemployment.[239]

Wilson Compton of the National Lumber Manufacturers Association agreed that the lumber industry needed to take a leadership role in conservation but also believed that lumbering needed to be self-sustaining. President Herbert Hoover responded to Compton's appeal for help in 1930, urging not outright government intervention but cooperation between public and private interests in what has been called the "associative state."[240] Hoover appointed the Timber Conservation Board to develop "sound and workable programs of private and public effort, with a view to securing and maintaining an economic balance between production and consumption of forest products." The board reported to the president a year and a half later with eighteen proposals.[241]

Besides the usual pleas for equitable taxation and better fire protection, the Timber Conservation Board recommended closer cooperation with private industry to coordinate public and private

The Civilian Conservation Corps put thousands of young men to work on diverse projects, such as building a bathhouse on the Sam Houston National Forest (right) and restoring totem poles in Alaska (facing page). (USDA Forest Service)

timber supplies and stabilize the marketplace. The board labeled this proposal to coordinate supplies "sustained yield." The term had traditionally meant that the volume of wood logged would not exceed the volume grown, but in this new use, the volume of wood was paramount, the quality of the wood was secondary, and other forest values received no consideration.[242]

The new definition of sustained yield, championed by industrial forester David T. Mason, a former Forest Service employee, had an economic component: public timber would be used to sustain lumber-dependent communities. When private supplies dropped too low to meet demand, public forests would be harvested while private lands regenerated. In this way, the government could coordinate yield with demand, stabilize the market, and foster conservation on private lands, all without direct regulation. Though Hoover ignored the report, the idea resurfaced in the New Deal.

Also during the Hoover administration, Pinchot's calls for significant increases in the amount of public ownership and management of forested lands began to find favor within the Forest Service. The idea appeared throughout the agency's report, *The National Plan for American Forestry*, more popularly known as the Copeland Report. Issued in 1933, the agency-wide effort was the first review of the activities and status of the national forests in more than a decade.

Produced in only a few months' time under the close direction and editorial control of Chief of Research Earle Clapp, the 1,677-page Copeland Report offered a comprehensive plan for more intensive management of all the National Forest System lands. The report evaluated virtually all aspects of forestry—public or private—including timber, water, range, recreation, state aid, and fire protection. In doing so, the Forest Service expressed the modern concept of multiple use in a substantive way for the first time. The agency saw no limitation to the simultaneous use of the land for grazing, recreation, and timber harvesting. As public tastes changed and Americans wanted to use the national forests for different activities, such as camping or hunting, the Forest Service believed that it could easily accommodate those activities. The report also included several proposals for conservation, such as reforestation and fire protection work.[243]

Not surprisingly, the Copeland Report praised the agency's own work as superior and blamed the major problems of American forestry on private owners. Eighty percent of the commercial forestland and ninety percent of the growth capacity were under private ownership, yet only ten percent of the "constructive effort" of managing forests, like building roads, was undertaken by private owners. The report suggested national planning through the massive public acquisition of private lands and public assumption of fully one-half of the work in forestry as the best solution. Showing restraint and flexibility, Clapp offered regulation of private cutting as a *quid pro quo* for public assistance and protection. The report also proposed significant public and private reforestation projects and greater cooperation for insect and disease control. Its comprehensive plan for more intensive management of the National Forest System lands would mean a more centralized federal administration and greater regulation of private lands.[244]

President Franklin Roosevelt at the first Civilian Conservation Corps camp on the George Washington National Forest in Virginia in 1933. (USDA Forest Service)

Clapp knew that some Forest Service officers bitterly opposed the fire protection recommendations, and others, the calls for regulation, but the report enjoyed limited congressional approval. Efforts to enact legislation based on its recommendations dragged on through the 1930s before Congress ultimately shelved them and the report with the outbreak of World War II. Meanwhile, both the lumber industry and the Forest Service sank deeper into the economic depression. Budget cuts and personnel layoffs loomed over the agency. The spread of white pine blister rust in the Northwest threatened to further destabilize the timber industry by flooding the market with diseased timber and driving down prices even more. The influence of the report, though, was substantial. It provided the blueprint for many of Franklin Roosevelt's New Deal conservation programs.

The inauguration of Franklin D. Roosevelt in 1933 brought a renewed presidential interest in land management and conservation. FDR had a lifelong interest in conservation and once listed his occupation as "tree farmer." His administration's accomplishments in conservation would rank second only to those of his fifth cousin and role model, Theodore Roosevelt. FDR turned for ideas to the Copeland Report, which had discussed hundreds of conservation projects requiring thousands of people to complete them. Through the newly created Office of Emergency Conservation Work, new programs put men to work building trails, picnic areas, and camps in the national forests. At Roosevelt's urging, Congress created the Civilian Conservation Corps (CCC) in May 1933 and called for a quarter-million paid volunteers to swarm the woods that summer to carry out conservation work. The U.S. Army, in charge of housing the men, placed one of the largest single orders in the history of the lumber industry; the available lumber glut made it easy to fill. Twenty-six hundred camps went up in a matter of months.

CCC funds also went toward purchasing cutover or abandoned lands, adding nearly 8 million acres to eastern national forests and ensuring that the men would have plenty of work during the CCC's decade of operation. Abandoned lands on the Great Plains totaling 11.3 million acres were also purchased for similar purposes; initially administered by the Soil Conservation Service, 3.8 million acres of these grasslands would later be placed under the control of the Forest Service. The Forest Service administered nearly half the CCC projects, from establishing trails and fighting forest fires to planting more than two billion trees and cutting hundreds of miles of firebreak in California's Sierra Mountains.

The Forest Service administered other New Deal conservation programs, such as the Prairie States Forestry Project, better known as the Shelterbelt Project. To address the ecological disaster known as the Dust Bowl, the Works Progress Administration provided the workers and the Forest Service the supervision to plant strips of trees at one-mile intervals on farmlands within a hundred-mile belt to break the prevailing winds that scoured the Great Plains between northern Texas and North Dakota. Under Forest Service leadership, 18,000 miles of shelterbelts provided protection to 30,223 farms while providing thousands of jobs to unemployed local workers. In 1937, FDR signed the Norris-Doxey Cooperative Farm Forestry Act to expand the Shelterbelt Project through a federal-state cooperative agreement to help farmers establish protective forests around their farms. In 1942, the shelterbelt work was transferred to the Soil Conservation Service.[245]

Grazing returned as an issue during the Dust Bowl years, in part because responsibility for the public range was divided among several agencies, including the Forest Service. Congress responded to the deteriorating range conditions with the Taylor Grazing Act of 1934, which created a grazing bureau in the Department of Interior to administer rangelands then under the control of the General Land Office. Representative Edward C. Taylor of Colorado, an opponent of federal controls, wrote the bill so that the proposed grazing bureau would fall under local domination. Most stockmen favored it for that reason; the Forest Service opposed it. The 80 million acres of grazing districts it created grew to 142 million acres shortly thereafter.

Raphael Zon convinced President Roosevelt of the need for a shelterbelt, a two-thousand-mile-long windbreak of trees planted between crop fields. It was designed to conserve topsoil and provide thousands of jobs while protecting more than thirty thousand farms. (USDA Forest Service)

Despite objections in Congress that the Shelterbelt Project would not work, the scientific feasibility of the idea was established more than thirty years earlier in Nebraska. In the 1890s, at the request of Division of Forestry Chief Bernhard Fernow, Charles Edwin Bessey, a professor of botany and horticulture at the University of Nebraska, conducted tree-planting experiments on private land in the Sandhills in north-central Nebraska. Within ten years, the planted pine trees were eighteen to twenty feet high. Gifford Pinchot, Fernow's successor, sent a survey team to examine the plantations in 1901. The team recommended the creation of two federal forest reserves on nearby federal land, and in April 1902, President Roosevelt created the Nebraska Forest Reserves. Foresters quickly established a tree nursery; planting commenced soon thereafter. In fact, they planted most of the forest, making it the only manmade national forest in the United States. In 1971, the federal government divided the Nebraska National Forest and created the Samuel R. McKelvie National Forest.

Source: Raymond J. Pool, "Fifty Years on the Nebraska National Forest," Nebraska History 34(3) (September 1953): 139–79.

The Forest Service responded two years later with *The Western Range*. The report was highly critical of Interior's handling of its range and of private range conditions but praised the management practices of the Forest Service. Issued in the middle of Interior Secretary Harold Ickes's attempts to have the Forest Service transferred to his department, the report served as justification for the status quo. Range protection was an agricultural objective, declared Secretary of Agriculture Henry Wallace, and the Forest Service was in his department.[246]

Ickes viewed the report as a thinly veiled threat and noted sarcastically that the Forest Service managed to turn over the six-hundred-page report only four days after a senator requested it. He had reason to be suspicious. At Wallace's request, the Forest Service had spent the previous four years investigating conditions, and it seized this opportunity to call for consolidating all federal range management under Agriculture. With the Grazing Service already established and the Great Depression deepening, however, President Roosevelt and Congress ignored the report. The issues remained unresolved for several years, and continued agitation by the Forest Service did little more than sap its own administrative energy.[247]

The New Deal programs opened the "intensive management" era of Forest Service history. The mysterious death of Chief Stuart in October 1933 after five years in office served as an unfortunate ending to "custodial management." Whether he fell from his office window or jumped has never been determined, but as with Henry Graves, the stress of the job had worn badly on Stuart. He took time off to rest in 1932 after a nervous breakdown, then returned to a worsening situation. The pressure of holding the agency together during the Depression and the intense mobilization efforts for CCC in spring 1933 added to the strain. At the time of his death, he was exhausted from leading the Forest Service during one of the darkest periods the nation has ever known. The strain would also kill his successor, Gus Silcox, who died of a heart attack, also after five years in office.

The Lumber Code

A few months before Gus Silcox became chief, Congress passed the National Industrial Recovery Act (NIRA), the capstone of FDR's first hundred days, a measure to which Roosevelt attached great importance and a reflection of the spirit of public and private cooperation characteristic of Hoover's policies. NIRA created the National Recovery Administration (NRA), which sought to establish national "fair competition codes" designed to eliminate unfair employment practices and cutthroat competition and revive the industrial economy.[248] FDR asked William Greeley, former Forest Service chief and now spokesman for the lumber industry, and David Mason, who had served on Hoover's Timber Conservation Board, to draft the lumber code. In setting minimum logging standards, Mason revived the Timber

Conservation Board's recommendation along with its definition of sustained yield. The Forest Service and the president requested and got stricter standards.

When the code went into effect in June 1934, midway through the life of NRA, operators were required to submit management plans to the government for approval. Greeley, Mason, and other industry leaders pushed the industry to accept the selection cutting technique written into the lumber code and supported by the Forest Service as the means to sustain yield. After NIRA and NRA were found unconstitutional eleven months later on the grounds that the federal government was exceeding its authority, the National Lumber Manufacturers Association announced that it would voluntarily retain the cutting principles in the NRA code.[249] Three-quarters of its affiliates signed on in agreement.

To the incoming chief, Gus Silcox, the lumber code presented an opportunity to regulate private industry and perhaps end destructive clearcutting on private land. Forest Service pronouncements from Washington emphasized the need for alternatives in the Northwest, an area targeted by the regulation crusade. But within the Forest Service, some researchers and field personnel disagreed with headquarters. They believed that selection cutting could encourage destructive insects and questioned the wisdom of leaving certain kinds of trees behind. Early Forest Service studies of clearcutting lent support to their position.[250]

The discussion became more confusing when it turned to the Forest Service's mission of multiple use. In light of the increase in recreational visits in the 1930s and the continued heavy

Frank G. Miller (left) and Charles A. Scott forded the North Platte River as part of the 1901 Sandhills reconnaissance party. Examinations of the area led to the establishment of the only manmade national forest in the country. (USDA Forest Service)

FORESTRY
—AND—
PERMANENT
PROSPERITY

U.S. DEPARTMENT OF AGRICULTURE
MISCELLANEOUS PUBLICATION No. 247

FORESTS
AND
FARMS

A GRAPHIC PRESENTATION OF THE
SOCIAL AND ECONOMIC SERVICES
OF THE
NATIONAL FORESTS OF CALIFORNIA

PREPARED BY
CALIFORNIA REGION, FOREST SERVICE
UNITED STATES DEPARTMENT OF AGRICULTURE
FOR THE
69TH NATIONAL GRANGE CONVENTION
SACRAMENTO, CALIF., NOVEMBER 13-21, 1935

The agency produced a steady stream of publications, such as these in the 1930s that emphasized the importance of timber management. One proclaimed, "National Forests are in reality huge timber farms, operated for the benefit of the Nation as a whole." (USDA Forest Service)

use by ranchers, Chief Silcox reminded the agency and the public in 1935 how the national forests were managed. "This multiple-use principle is not new," he announced. "It has been applied to the national forests for more than 30 years." It was, though, just that—a principle. There was no law, no plan, and no set of regulations that gave equal consideration to the various uses. The Forest Service made management decisions based on what was best for timber. Where timber dominated, "sustained yield" was the watchword. Where there was little or no timber, multiple-use management was easier.[251]

National forests in Oregon had examples of both sustained yield and multiple-use areas. In the diversified economy of Umatilla County, foresters embraced multiple-use planning to ensure "continuous production of the various forest contributions such as lumber, forage, watershed protection, and recreation."[252] But in Multnomah County, home to timber-dependent communities, it was sustained yield that would ensure economic health. In the 1930s, the goal shifted from liquidating mature timber as fast as possible to cutting at a slower rate that would produce a second harvest as large as the first one. In several instances, however, the Forest Service turned a blind eye to the overcutting of public timber to keep mills open, but overcutting also dumped lumber into a flooded market.[253]

The Forest Service campaign for federal regulation meanwhile divided membership of the Society of American Foresters along private and public forestry lines. With its own timber sales low, the Forest Service could hope to control logging practices only on other people's land. Industry countered by getting Congress in 1937 to authorize the establishment of sustained yield as a principal management tool using private and Interior lands in the Northwest, an idea based on David Mason's concept of sustained yield. By combining timber from private and public lands, lumber companies would not have to move their mills, and lumber-dependent communities would remain healthy.

Forest Service opposition to that move was undercut by its own figures showing that there was more timber than had been previously estimated, especially in the Northwest; that regeneration had kept pace with the boom harvests of the 1920s; and that private companies were indeed employing conservative logging methods. Chastened but not repentant, the Forest Service established its own administrative policy: lumber interests cutting federal timber had to practice sustained yield and "must agree" to meet minimum standards on their own land.[254]

In the buildup to World War II, Forest Service reports and statements continued emphasizing selection harvesting and criticizing clearcutting, but the language was often at odds with actions. The agency was pleased to observe sustained-yield harvesting on industrial lands in other parts of the nation, particularly the South, but the Northwest, where most of the timber was, continued to draw its ire. Industry countered regulation efforts in 1941 by forming the Forest Farmers Association in the South and the American Tree Farm System in the Pacific Northwest; the latter eventually became a national organization. Both organizations lobbied Washington on behalf of private timberland owners.

Angered by the establishment of the Tree Farm System, Chief Lyle Watts, who championed regulation while serving as regional forester in the Pacific Northwest before becoming chief in 1943, declared he wanted to "kill the thing before it spreads." The organizations won a major victory when Congress overrode a presidential veto and passed a 1944 federal income tax law that authorized lumbermen to report income from timber sales as capital gains instead of conventional income.[255] It thus encouraged landowners to hold on to their land and implement good management practices. Undeterred, the agency pursued federal control of private lands and flogged industry for clearcutting Douglas-fir, even though Forest Service research supported and agency timber managers encouraged that silviculture method. This tactic, however disingenuous, made it easy for the Forest Service to draw distinctions between itself and private industry while reminding the public of the continued need for the Forest Service.

Forest Farms

The Forest Service's constant calls for regulating private timberlands motivated private landowners in 1941 to organize the southern-oriented Forest Farmers Association and the nationally focused American Tree Farm System. Calling themselves tree farmers made sense. In 1935, Gifford Pinchot had declared in a speech, "Wood is a crop. Forestry is Tree Farming." Both organizations formed to disseminate information and ideas about forestry practices but also to promote their political interests, particularly tax reform. Because it penalized timberland owners for holding on to the land, the existing tax system fueled cut-and-run logging: it was cheaper to purchase new land than to pay taxes on one's original property. Both organizations also pushed for cooperative fire protection and research with state and federal agencies and promoted good management practices among its members. As more and more timber has come from private lands in the past few decades, these two organizations have grown in size and importance.

Source: Lester A. DeCoster, "Tree Farming Tenacity: Sixty Years Old and Still Going Strong," Tree Farmer (July–August 2001): 6–15; and Forest Farmer, "Forest Farm Association through the Years," Forest Farmer (May 1991): 18–47.

World War II placed extraordinary demand on the Forest Service for natural resources. The federal government used posters and print ads to encourage the public to conserve rubber and naval stores. (USDA Forest Service)

War Demands

Though the Forest Service disagreed with Mason's idea of sustained yield because it made federal regulation on private land unnecessary, Congress, private industry, and the enormous wartime demand for lumber left the agency with few options. Military construction and exports boosted consumption from five billion board feet in 1941 to more than twelve billion the following year, and an increasing amount of that came from national forests as private timberland stocks dwindled.

The lumber became barracks, warehouses, hangars, shipyards, and portable docks called mulberry harbors for the Normandy invasion. Wood in solid, laminated, and plywood forms was a basic structural component in landing vessels, PT boats, patrol boats, and mine sweepers and provided decking on aircraft carriers and battleships. Wooden packing crates protected tons of war materials being shipped overseas, and wood was used in munitions, aircraft, and trucks. The Forest Products Laboratory in Wisconsin conducted much of the research on wood use during the war.[256]

FOREST DEFENSE IS NATIONAL DEFENSE

Reprinted from The American Weekly, April 6, 1941

HOW TO PREVENT FOREST FIRES

FOREST FIRES AID THE ENEMY

Crush out your cigarette

U. S. DEPT. AGRICULTURE, FOREST SERVICE   STATE FOREST SERVICE

When the war in the Pacific cut off rubber supplies to the United States, Forest Service scientists found a viable substitute in a southwestern shrub, the guayule plant. Within three months, the project had a fifty-four-acre nursery to produce seed, a fifteen-hundred-man labor camp, a seed extractory, a rubber extraction plant, two laboratories, and an equipment design and repair shop. Impressed, Congress increased funding to bring more acreage under cultivation, and by spring 1944, researchers had more than two hundred thousand acres under cultivation. The Forest Service produced three million pounds of the substitute and developed a process for mixing it with natural rubber to make leak-proof fuel tanks for military aircraft.[257]

As the war drained the Forest Service of men, the agency hired women to protect the national forests. The Forest Service had its own version of Rosie the Riveter. The Shasta National Forest (today the Shasta-Trinity) called their women workers Shasta Susies. The Portland regional office hired 246 women for the 1943 fire season in Oregon and Washington to fill fire protection jobs, which included lookout service as well as fire dispatchers, cooks, telephone operators, and clerks. Women patrolled campgrounds and worked as truck drivers. They served as aircraft observers in the Aircraft Warning Service (AWS), scanning the skies for enemy planes. From 1942 to early 1943, when coastal radar came online and rendered AWS redundant, they kept a 24-hour,

Part of the war effort involved increasing public awareness of the dangers of forest fires. The agency's firefighting units had dwindled to skeleton crews because of the loss of personnel to the military. Preventing fires was the Forest Service's best weapon. (USDA Forest Service)

Women's volunteer motor corps at the ready in the Trinity (now Trinity-Shasta) National Forest in 1943. The Forest Service relied on women to fill in for men called up for military service during World War II. (USDA Forest Service)

365-day vigil, reporting each plane spotted, the type of aircraft, and its direction.[258]

Other women took over the duties of rangers and supervisors without pay. The head of the Santa Barbara Red Cross volunteered her "good horsewomen" to patrol the Los Padres National Forest. She told the supervisor, Stephen Nash-Boulden, not to pay the women. Nash-Boulden worked out a plan in which each woman donated her salary to the Red Cross, and for the duration of the war, Red Cross women patrolled the forest.

The agency also called upon the military and the Civilian Public Service for help in carrying out vital duties. The U.S. Army had adapted several ideas for its paratroopers from the Forest Service smokejumpers, the elite firefighters who parachuted with their supplies to fight remote fires. When Japan started sending balloon-borne incendiary bombs to ignite west coast forests in spring 1945, the Army responded with Operation Firefly. It dispatched the first all-black battalion of paratroopers, the 555th Parachute Infantry Battalion, the "Triple Nickles," along with some conscientious objectors to serve as smokejumpers in the Pacific Northwest and northern California. The Triple Nickles underwent three weeks of intensive training, including demolition of unexploded ordnance, firefighting, wilderness survival, and first-aid. Refitted with new gear—football helmets with wire masks that protected the jumpers' faces as they dropped through trees and fleece-lined flying jackets and trousers instead of the usual canvas jumpsuits—the men made three practice jumps before being dispatched to their bases in Pendleton, Oregon, and Chico, California. From mid-July to early October 1945, the Triple Nickles participated in thirty-six missions and made twelve hundred jumps; only one man died in action. After the war, the battalion was sent back to North Carolina and absorbed into other airborne units.[259]

The Civilian Public Service offered "work of national importance" to conscientious objectors during the war. Under its auspices, about twelve thousand of the war's seventy-two thousand conscientious objectors continued CCC work by fighting fires, planting trees, and conducting watershed development for the Forest Service, the National Park Service, and the General Land Office. During two years on the Modoc National Forest in northern California, for example, they planted almost 748,000 Jeffrey and ponderosa pine seedlings on about twelve hundred burned-over acres. A few hundred more adventurous men became smokejumpers and made about five thousand jumps in three seasons. Some of the smoke-jumpers also trained military personnel for rescues.[260]

Sustained-Yield
Forest Management Act of 1944

Throughout the war, timber production remained the most important job of the Forest Service. In 1944, Congress passed the Sustained-Yield Forest Management Act, which industrial forester David T. Mason helped draft. The law, similar to the plan Mason had devised for the Department of Interior, permitted the Forest Service and lumber companies to enter into long-term agreements that promised a constant supply of public timber at or above appraised value, without competitive bids. There were two types of contracts—cooperative sustained-yield contracts and federal sustained-yield contracts. This law guaranteed supply only when a community's stability left it no choice but to cut federal timber not otherwise available through conventional sales. Proponents argued it was cheaper to give up revenues lost from noncompetitive timber sales than to place entire communities on the welfare rolls. The argument convinced Congress. With its passage, the greatest good expanded to include sustaining timber-dependent communities.

Two days after FDR signed the bill on March 29, 1944, the Simpson Logging Company of Shelton, Washington, filed the first application for a cooperative agreement to prevent dismissing five thousand of its six thousand workers in the next few years. The Simpson deal combined federal with private lands for a total of 270,000 acres of forestland; eighty-two percent of the standing timber was on federal lands. The agreement prevented economic catastrophe, and federal controls safeguarded the public interest: the legislation worked as designed.[261]

Just because it worked did not mean the law proved popular, however. In the case of Simpson Logging, the two parties entered into a hundred-year contract in December 1946. Neighboring towns that were not going to benefit and companies ineligible to bid on timber from the national forests complained that the deal granted Simpson a monopoly over federal timber and restricted other uses of the land, including farming and recreation; industry saw

Jesse Mayes was the executive officer of B-Company of the 555th Parachute Infantry Battalion. The Triple Nickles made twelve hundred jumps during Operation Firefly.
(Missoula Smokejumper Visitor Center, USDA Forest Service)

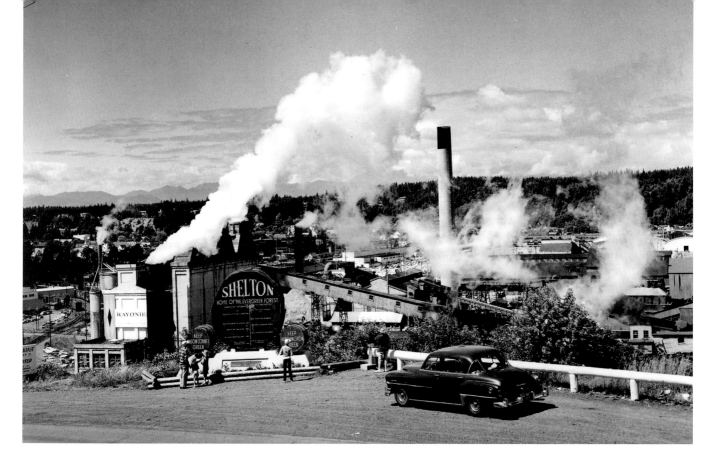

For one timber-dependent town in Washington, the Shelton Cooperative Sustained-Yield Unit meant that the Simpson Logging Company would not have to shut down operations. So much controversy surrounded the deal that the federal government made no other public-private agreements like it. (USDA Forest Service)

the Simpson contract as a new way for the Forest Service to regulate logging on private lands.[262] The controversy was such that the Forest Service made no further cooperative sustained-yield units. However, it did enter into five agreements that used only federal land. The agency reserved a total of 1.7 million acres to stabilize five communities in New Mexico, Arizona, California, Washington, and Oregon. Other proposals, denounced by industry as communistic, socialistic, and un-American, were dropped.

Lyle Watts, the Forest Service chief who approved the Simpson deal, believed that support from the local community proved that the cooperative idea succeeded. To Watts, though, success meant that the Forest Service retained control, even if indirect, over logging. Ignoring the complaints from neighboring communities, he concluded that Shelton's interests represented those of the larger public and that he could legitimately dismiss opponents who favored their own interests.[263] This insistence on doing it the Forest Service way was not unique to Watts. It became the *modus operandi* for the Forest Service for the next several decades: timber management on public lands would be carried out as the Forest Service saw fit, not on narrow interpretations of the greatest good.

As chief from 1943 to 1952, Watts continued pressing for federal regulation. Until this point, everything had depended on the Forest Service's ability to convince the public and Congress that the agency alone stood between the lumber companies and timber famine. But more than twenty years of fighting over the issue of regulation had cost the Forest Service valuable congressional support by the time Watts took charge. Under Watts, support eroded further and the agency lost important backing for fire control and other measures. The issue divided the agency as more field personnel who had to live in or near the lumbering communities found themselves siding with industry. After the war, it appeared that continued calls for regulation might hinder efforts to meet postwar lumber demand and prevent loggers and others in the industry from prospering.[264]

By the 1940s, the debate over regulation had changed. The technical issue was largely resolving itself, and without federal intervention. Earlier reforestation efforts in the South meant that by the 1930s, timberlands in many areas were ready for harvesting. Creation of the tree farm system demonstrated that many in private industry were moving toward sustainable forestry practices without federal mandate.

Pinchot and other foresters of his generation, trained to see timber supply in terms of physical volumes, had neglected to take into account economic and social factors. Looking only at trees on the stump and the rate at which forests were being cut in the early 1900s, they concluded that the nation would run out of timber. What Pinchot failed to notice, according to forest policy analyst Sally Fairfax, was that timber consumption was "an intersection of supply and price; it's an economic matter. If scarcity takes hold, the price will go up or substitutes will be found." Changing technology had also affected timber consumption. Train engines no longer burned wood, and because of wood preservation technologies, railroads had reduced their use of crossties. The automobile led to a drop in rail traffic. Home heating had shifted from wood-burning stoves to coal to gas and oil. After the war, the use of herbicides and new silvicultural methods increased the growth rates of trees and the volume of timber produced. Technology and science had reduced the likelihood of a timber famine, even if Pinchot did not realize it.[265]

The philosophical side of the debate had raged ever since 1919, when Pinchot proposed regulation that opponents had labeled socialistic. In the early years of the Cold War, which pitted the democratic, free-market economic system led by the United States against the socialistic, government-controlled system led by the Soviet Union, equating public ownership with socialism doomed regulation. The Korean War, a struggle against communism that began in 1950, placed more pressure on the national forests for timber while further demonizing socialism.

The regulation debate ended with the appointment of Richard McArdle as chief in 1952 and the election of war hero Dwight Eisenhower as president that same year. The Eisenhower administration had promised to roll back many of the New Deal's regulatory efforts and had no interest in letting the debate continue. McArdle concurred. With that issue resolved, the Forest Service and private industry began to mend fences and work together to meet rising timber demands.

Chapter Five

RECREATION, WILDERNESS, AND WILDLIFE MANAGEMENT IN THE FIRST HALF-CENTURY

Summer cottages on the Angeles National Forest, 1916. (USDA Forest Service)

UNTIL THE 1940S, TIMBER PRODUCTION IN THE NATIONAL FORESTS WAS A PRIORITY, EVEN THOUGH NOT MUCH LOGGING ACTUALLY OCCURRED. AS PART OF ITS CUSTODIAL MANAGEMENT DUTIES, IN THE 1910S THE FOREST SERVICE BEGAN WORKING TO RESTORE WILDLIFE AND GAME POPULATIONS TO ATTRACT SPORTSMEN TO THE NATIONAL FORESTS. AS RECREATIONAL USE OF PUBLIC LAND INCREASED AFTER WORLD WAR I, NEW CONCEPTS OF WILDERNESS AND HOW THE PUBLIC USED THE NATIONAL FORESTS EMERGED. THE FOREST SERVICE RESPONDED BY DESIGNATING WILDERNESS AND PRIMITIVE AREAS IN THE 1920S. RECREATIONAL USE INCREASED AGAIN IN THE LATE 1940S AT THE SAME TIME THAT RESOURCE EXTRACTION AND LOGGING TOOK OFF. WITH THE INCREASE CAME PRESSURE FROM RECREATIONAL USERS FOR THE FOREST SERVICE TO PROVIDE PERMANENT PROTECTION FOR AREAS THEY FAVORED. UNTIL THE LATE 1950S, HOWEVER, THERE WAS NO SYSTEMATIC APPROACH TO MANAGING ALL OF THESE USES OR ACCOMMODATING CONFLICTING INTERESTS. THE DEMAND FOR MULTIPLE USES OF THE LAND BROUGHT ABOUT A NEED TO REARRANGE PRIORITIES.[266]

Hetch Hetchy Valley

Before automobiles became commonplace, the national forests served the needs of locals, who used them chiefly for sustenance. The handful of national parks and national monuments then in existence had been established to protect and celebrate unique visual assets—the sweeping grandeur of the Grand Canyon, the otherworldly hot springs and geysers of Yellowstone Park in Wyoming, and the majestic views of Yosemite Valley in California. Congress did not establish new parks unless assured the land had no economic value beyond recreation and aesthetics.

National parks and monuments fell under the jurisdiction of the department controlling the land. The Department of Agriculture had twenty-one monuments under its jurisdiction, and many parks were surrounded by national forests. Chronically shorthanded, the Forest Service could do little to protect or manage the sites, leaving them open to theft and vandalism. Nonetheless, as chief forester, Gifford Pinchot targeted those parks and monuments not controlled by the Forest Service for transfer to his agency—unsuccessfully, however.

Where competing interests intersected, confrontations soon followed, and forest recreational use and management policy developed along two paths. Preservationists, like John Muir and those who advocated for Yosemite National Park, conceived of the public forest as a natural setting, best represented by national and state parks, in which the forest served as a background or space for various activities and experiences. They made their appeals for preservation to the

Understaffed and underfunded, the Forest Service had difficulty in the early years protecting historical sites like these cliff dwellings and drawings found in the Santa Fe National Forest in New Mexico. Consequently, preservationists demanded a separate agency to manage national parks and monuments. (USDA Forest Service)

public's emotions. The second path, as first articulated by Gifford Pinchot as "the greatest good" but fully implemented by Graves and others, stressed utilitarianism—all the assets of the forest are for use, and recreation is just one of many uses. Muir and his followers led with their hearts; Pinchot and his acolytes led with their heads.[267] The early Forest Service chiefs favored timber and commercial use and considered recreation a compatible commodity use. In practice, though, recreation took a backseat to timber management. Timber was driving the bus in which the recreating public was riding.

The debate over how to define the greatest good for the greatest number on federal land in California led to the first significant national debate between preservationists and conservationists. The Forest Service, as outlined in the 1897 Organic Act, could engage in timber production to protect watersheds, which sometimes surrounded parks and national monuments. The policy left the agency vulnerable to criticism. In 1905, Congress transferred 542 square miles from Yosemite National Park to the adjacent Sierra Forest Reserve because it deemed the land too valuable for lumbering, mining, and grazing. The move put preservationists on notice. Shortly afterward, the city of San Francisco set its sights on Hetch Hetchy Valley in Yosemite National Park to meet its increasing municipal water supply needs. John Muir, founder of the Sierra Club and protector of the park, vociferously opposed the plan to build a dam there and squared off against his friend Gifford Pinchot, who sided with San Francisco, as did the Roosevelt administration. The Sierra Club did not want to preserve Hetch Hetchy as untrammeled wilderness; it wanted to prevent the valley from being flooded so that it could be used for recreational tourism. The ensuing debate brought national attention to the divisions within the conservation movement, made the Forest Service appear hostile to the concept of national parks, and added to the image of a power-hungry Pinchot. After several years, President Woodrow Wilson signed legislation in 1913 permitting the dam's construction.[268]

The schism, however, was not just between two friends. It was between two schools of philosophy in land management. Preservationists believed they could no longer trust the Forest

The dam at Hetch Hetchy Valley took twenty years to construct and cost $100 million, twice as much as originally expected. Debate over the dam's location pitted preservationists against conservationists and led to the establishment of the National Park Service. (Top photo by Joseph N. LeConte – Sierra Club Library Collection; Bottom photo from author's personal collection)

Service to protect national parks in perpetuity from commodity development, even though Pinchot opposed logging in Yosemite. Organizations like the Sierra Club pushed for the creation of a separate federal agency to manage the national parks for recreational tourism. Forest Service leaders argued that they, too, valued recreation and beauty and, furthermore, that they were better qualified for managing both parks and forests. To preservationists, however, Forest Service

John Muir escorted Presidents Theodore Roosevelt (on left, with Muir) and William Howard Taft through Hetch Hetchy Valley in Yosemite in an effort to persuade them to save it. Muir found Roosevelt willing to camp with him but unwilling to listen, and Taft willing to listen but unwilling to camp.
(Library of Congress)

support of a dam in Hetch Hetchy when other sites were still available provided evidence against the agency's case.

An Unfriendly Rivalry

As early as 1907, while dean of forestry at Yale, Henry Graves argued that the national parks belonged in the Agriculture Department because of the need to salvage the dead and down timber. As forestry chief, he continued reasoning along those same lines, contending there was only a slight difference between parks and national forests, the difference being purpose, not administration. Inside the parks, he said, the Forest Service would curtail the extent of timber cutting for commercial purposes, but both types of reserves required technical foresters.[269] Graves remained puzzled by the public disapproval of his proposal for logging under controlled conditions in national parks.

Graves was not hostile toward parks, as many preservationists thought, but he saw no need for a separate national park bureau. Given the Forest Service's limited manpower and budgets, however, grazing and timber were higher priorities than recreation. Nor was there much obvious public pressure for recreational facilities in the national forests. In fact, much of the public ranked timber, fire control, grazing, and other responsibilities above recreation. In the early years, fish and game management to propagate species for fishing and hunting, along with building roads, issuing summer home permits, and marking trails for public use, constituted the extent of the Forest Service's efforts to support recreation.

After Hetch Hetchy, the Forest Service altered its stance on recreation in hopes of staving off creation of a separate park bureau. Herbert Smith, who handled the agency's public relations, touted the recreational uses of the national forests in press releases. In 1915, the Forest Service secured the Term Permit Act so that it could offer recreation residences in the national forests and curried favor with the influential people who leased these summer homes. The agency opened its first campground in 1916 at Eagle Creek on the Oregon side of the Columbia River Gorge on the Mount Hood National Forest. But these attempts to stop the parks movement proved too little, too late. Recreational use of the forest was an issue of the heart, not the head.[270]

President Wilson handed preservationists a victory when he signed legislation creating the National Park Service to manage national parks and monuments in August 1916. Its first director, Stephen Mather, a prominent Sierra Club member and critic of the Forest Service's handling of the forests, had it up and running early the next year. Mather, a millionaire who had made his money from borax mining, wrote to Secretary of Interior Franklin K. Lane to protest the deterioration and poor handling of the national parks after a camping trip to the High Sierra in 1914. At the time, Lane was looking for a way to offset the political damage suffered for backing the Hetch Hetchy project, which, as the city attorney for San Francisco a decade earlier, he had advocated. Mather, a member of the Sierra Club since 1905, had campaigned against the dam. To bring the parks under systematic management and simultaneously repair the damage his support of Hetch Hetchy had caused, Lane appointed Mather as his special assistant for national park concerns in January 1915, and then as director of the National Park Service in May 1917.

As the "back to nature" movement began to take off in the interwar period, Mather sought to capitalize on interest in visiting national parks as a way to strengthen his bureau's standing. In Mather's vision, parks would be made accessible by automobiles and trains and developed for public sightseeing, education, and related uses. Neither Mather nor the Forest Service anticipated that opening up the lands for tourism would contribute to the degradation of the land and actually put the areas at risk.[271]

Even after Mather took charge of the Park Service, Graves continued to question the logic of having two agencies in different departments with overlapping responsibilities. He championed governmental efficiency, and had a lingering distrust of the ability of the Department of Interior's General Land Office to manage land. By lumping together camping, fishing, and hunting as recreation, the Forest Service could argue that the Park Service provided only a single use of

John Muir on the Preservation and Use of Forests

John Muir, founder of the Sierra Club, has been frequently depicted as seeking to cordon off forests as wilderness areas or as opposing the mission of the Forest Service. On the contrary, Muir recognized that the national forests were for use, and proclaimed they should be managed for that purpose thus:

[I]t is impossible, in the nature of things, to stop at preservation. The forests must be, and will be, not only preserved, but used; and the experience of all civilized countries that have faced and solved the question shows that, over and above all expenses of management under trained officers, the forests, like perennial fountains, may be made to yield a sure harvest of timber, while at the same time all their far-reaching beneficent uses may be maintained unimpaired.

Source: John Muir, "A Plan to Save the Forests: Forest Preservation by Military Control," The Century Magazine 49(4) (1 February 1895): 631.

Summer recreational homes in national forests ranged in style from one-room cabins to elaborate homes owned by influential people, including publishing magnate William Randolph Hearst, whose home is shown here in 1914. (USDA Forest Service)

the land. The Forest Service, in contrast, allowed recreation along with extractive activities—multiple uses—on its land. It was a difference in semantics, but one the Forest Service embraced for years as proof of its multiple-use strategy and the Park Service's duplication of its work.[272]

Whether the Forest Service increased its recreational programs in response to creation of the Park Service is not clear, but it appears more than a little coincidental that in 1917, one of the foremost landscape specialists in the country, Frank A. Waugh, completed a thorough study of Forest Service recreation facilities. Waugh spent several months inspecting national forests, inventorying recreational resources and recreational activities in campgrounds, picnic areas, and summer homes, then developed plans for new facilities that could compete directly with what the Park Service had to offer. After visiting the Grand Canyon, which was still under Forest Service control in 1917, Waugh generated a plan to accommodate a thousand visitors a day in comfortable fashion. It included a dozen hotels, thousands of cottages, a tramway to the canyon floor, and a $100,000 water development project. Plans like this demonstrated that the Forest Service shared Mather's "people's playground" philosophy while boosting the agency's contention that it could do it all.[273]

Waugh tried to establish an economic value for recreation in his report. Putting a price tag on nature's aesthetic value, he calculated that national forest recreation was worth a minimum of $3 million annually, a value much greater than its costs: "The National Forests are certainly a paying investment for the American people."[274] Graves supported this position in his own sixteen-page policy statement issued two years later. He saw the park system as both a challenge to and a disruption of the National Forest System and argued that critics of the Forest Service misrepresented the agency's attitudes toward recreation and parks. Natural resources in parks should remain accessible, he believed, so that the Forest Service could tap them if necessary. Until then, they would remain fully protected.[275]

Chief William Greeley's first annual report, in 1920, marks a noticeable shift in the agency's approach to profit. He maintained the long-standing contention that forestry was profitable, but he now included language about the nonmonetary benefits of forests, declaring, "In addition to this revenue, there is an enormous return to the public through…the recreational facilities made available to hundreds of thousands of our people. The monetary income from the National Forests can be expected to increase steadily. But there will always be national returns not measurable in dollars which in public benefit exceed the receipts paid into the Treasury."[276] He then recounted the progress made in recreation and game management, and how much money from user fees each returned to the Treasury. To this utilitarian forester, at least, nonmonetary benefits could return a profit after all.

Tension between the two agencies continued after Greeley became Forest Service chief. Creating more national parks typically meant carving them out of existing national forests, and thus the Forest Service was literally losing ground to the Park Service. A Forest Service map of California's national forests left the national parks as blank spaces—an inadvertent salvo fired in what Greeley in 1921 called the "continued warfare."[277] Chief Greeley met with a furious Director Mather in an attempt to improve relations and agreed to jointly examine national forest lands under consideration for national parks. The Forest Service would hand over to the Park Service land "where the dominant resource consists of scenic features of such a character as to have national importance," but "areas whose dominant resources are economic or whose scenic and recreation features are not of outstanding importance should remain" as is. Forest Service personnel agreed to give the Park Service credit for its programs when speaking about recreation. Furthermore, the Forest Service would be considerate of the Park Service's mission when it permitted logging and grazing near parks.[278]

Ironically, neither agency had a clearly articulated plan for recreation or how to use unimproved areas.[279] The need for plans to serve the public's growing recreational needs became more acute in the 1920s as mass production of automobiles, expansion of the national highway system, and longer vacations greatly increased the demand for multiple-use recreation areas throughout the nation.[280] The Park Service had already begun building more roads inside the parks to open up more areas to visit. To study the situation, President Calvin Coolidge convened the National Conference on Outdoor Recreation in 1924, which brought together 309 delegates from 128

The first recreation boom came during the interwar period. Scenes of overcrowded campsites like this one on the San Bernardino National Forest, California, in July 1938 foreshadowed what was to come after World War II. (USDA Forest Service)

organizations. The conference appointed a joint committee of the American Forestry Association and National Park Association to survey federal parks, forests, and other public lands and waterways.

The 1928 joint committee report strongly recommended creating roadless wilderness areas to provide a different kind of recreation experience from that of the automobile user. In support of this position, the committee excerpted a 1925 article by Forest Service employee and wilderness advocate Aldo Leopold, who called for preserving wild areas for unique recreational opportunities before all significant wilderness disappeared.[281] The report also discussed the many meanings of recreation on public lands, from day trips and scenic drives to picnicking and weekend car-camping in organized campgrounds to month-long backcountry canoe trips.[282] At the heart of debate over recreation was the place of the automobile and its impact on land use.

Automobiles did more to alter national forests and the work of the Forest Service in its first twenty years than anything other than fire. During those years, the Forest Service constructed roads largely for fire suppression. Those roads also opened up the national forests for recreational use, and dealing with summer homes, resorts, and family camping occupied increasing amounts of Forest Service time.

Transfer Threats

In addition to fighting over the land itself, the Departments of Interior and Agriculture fought over control of the agencies themselves. Shortly after taking office in 1921, Secretary of Interior

Albert Fall attempted to have the Forest Service transferred to his department. When that failed, he proposed that the National Park Service administer recreation on national forests or, as an alternative, that national monuments under Forest Service administration be transferred. Fall's involvement in the Teapot Dome scandal that rocked President Warren G. Harding's administration put a hold on the transfer idea, however. For their part, Forest Service and Department of Agriculture officials made three attempts in the 1920s and early 1930s to have the Park Service transferred to the Department of Agriculture and placed under Forest Service jurisdiction. Disagreements over park and forest boundaries came from both bureaus, which further heightened tensions.[283]

Franklin Roosevelt's New Deal conservation programs touched off the greatest challenge to Forest Service prerogatives yet. Tensions boiled over when Secretary of Interior Harold Ickes, a longtime conservationist, took office in 1933 with President Franklin Roosevelt's support for many of his ideas. In June 1933, Roosevelt signed an executive order immediately transferring control of all national monuments to the National Park Service, which meant the Forest Service lost control of fifteen monuments. Inexplicably, the Forest Service meekly protested the transfer. Though it made administrative sense, Forest Service pride still suffered a major blow. Ickes then set about increasing his department's responsibilities and land holdings by creating more wildlife refuges, marine sanctuaries, and other areas of wildlife protection. When he got wind of Forest Service plans to authorize logging in the Olympic National Forest on land adjacent to the Olympic National Monument, Ickes and his supporters pushed through a bill incorporating the land into a national park, and thus valuable timberland was transferred from the Forest Service to the Department of Interior.[284]

Roosevelt also supported Ickes's idea of consolidating all public land management agencies, including the Forest Service, into a new Department of Conservation. Forest Service leaders, outraged by the idea of becoming associated directly with what they considered Interior's disgraceful past, called on former chiefs, along with the American Forestry Association and the Society of American Foresters, to lobby Congress and the president. Nearly every organization outside government concerned with natural resources—with the notable exception of the livestock industry, which was feuding with the Forest Service over grazing limits—came out in opposition to the Interior Department's proposed takeover of the national forests.

The debate continued during the rest of Roosevelt's first administration and erupted into a melee in 1937. An executive committee on reorganization of the executive branch recommended transfer of the Forest Service to Interior. The mudslinging that ensued got personal and destroyed the longtime friendship of Ickes and Pinchot. Ickes accused Pinchot, a lifelong Republican who had served two terms as Pennsylvania's governor and had never hidden his presidential aspirations, of siding with "a motley crew of lumber barons" in his attempt to get a cabinet seat or even run for the White House. Pinchot, by now in his early 70s, still had plenty of fight in

According to Forest Service lore, Gifford Pinchot left behind a box containing secret instructions on how to stop any attempt to transfer the Forest Service to the Interior Department. Former Chief Jack Ward Thomas looked for it,

...but I could never find it. On the other hand, if I had, I wouldn't tell you, obviously. But, the Black Box, theoretically, is to be opened when the Forest Service is in danger of being transferred to the Department of the Interior and it's the plan on how to stop that transfer. Well, maybe the reason I never found the Black Box is that Gifford's ghost didn't reveal it to me because on my watch there was no danger of this happening. I'm sure, if need be, Gifford will appear to some future chief and show him where the Black Box is. But there's an equivalent [of the] Black Box that was in Pinchot's desk, which says, he who sits in this chair needs to come to work every day prepared to be fired. Maybe *that's* the Black Box, I don't know. It was never revealed to me, but I'm sure that it exists somewhere and that at an appropriate time, Gifford's ghost will reveal it to a Forest Service chief sometime in the future.

Source: Jack Ward Thomas, interview for The Greatest Good: A Centennial Film.

him. Always eager to defend his former agency against outside critics, Pinchot compared Ickes to German dictator Adolph Hitler when he declared, "Grabbing for power is not well regarded in the world of today." Greeley, also caught up in the wartime rhetoric, compared the transfer to the dismemberment of Poland, saying it would leave fragments of the Forest Service "strewn among the other agencies."[285]

The strength of the opposition made it clear to Roosevelt, already facing accusations of making a power grab of his own by "packing" the Supreme Court, that the transfer made little political sense. Though the issue was largely settled by 1938, the wounds lingered. When Chief Gus Silcox died of a heart attack in 1939, FDR named Earle Clapp acting chief but refused to make the appointment permanent once he learned of Clapp's opposition to the transfer. Transfer rumors lingered for a few more years and subsided only when the nation went to war in 1941. In 1943, Lyle Watts was named chief to replace Clapp.

The postwar years brought yet another threat of transfer. Until World War II, grazing was the largest use of public land, taking place on 100 of the 152 national forests. As with timber, the world wars had demanded full use of public ranges. After World War II, the Forest Service wanted to reduce the head counts and restore overgrazed lands; the land could then provide other uses. But the dramatic rise in demand for beef led ranchers to demand that the Forest Service allow more grazing on long-term permits. In the midst of this debate, President Harry Truman appointed the Hoover Commission in 1948 to review governmental organization and identify overlapping activities that could be consolidated or eliminated. The Forest Service appealed to its conservation allies for help in turning back the ranching lobbies and averting any new possibility of transfer by invoking multiple use. It was one of the last times the Forest Service and the preservationists saw eye to eye on land management.[286]

Wilderness and Recreation

The tensions between the National Park Service and the Forest Service in the 1920s and 1930s pushed the Forest Service to rethink its policies regarding recreation and wilderness and to move more aggressively toward managing its lands for multiple uses. During the early debates about recreation, Chief Greeley and many in the Forest Service recognized the validity of preserving and setting aside wilderness areas for recreational purposes. Greeley had been addressing the topic of wilderness and recreation within the Forest Service when the National Conference on Outdoor Recreation convened in 1924. He favored prohibiting roads and summer homes in wilderness areas and believed access should be by foot, horseback, or canoe. The Forest Service, though, had no tally of how much wilderness remained on national forests, and Greeley wanted to retain administrative flexibility. Declaring large undeveloped areas of national forests off-limits did not fit that model. In 1926, the agency began a national

inventory of all areas greater than 230,400 acres to determine the extent of wilderness remaining on national forests and reported that seventy-four tracts totaling fifty-five million acres remained.

Based on that information, the Forest Service announced two additional wilderness designations in 1929: research reserves and primitive areas. A research reserve was an area preserved for scientific and educational purposes. Under administrative regulation L-20, regional offices nominated minimally developed primitive areas, where visitors could get a sense of conditions typical of the pioneer era. This did not mean, however, that timber production, grazing, or mining were permanently excluded. Of the seventy-two primitive areas comprising a gross area of 13.5 million acres in ten western states, the Forest Service allowed road construction in fifteen areas, grazing in sixty-two, and logging in fifty-nine. Only four primitive areas totaling 297,221 acres absolutely excluded those activities.[287]

Even if born of noble intentions, a wilderness policy nonetheless had to be reconciled with the other obligations of national forest administration. Setting aside wilderness areas, for example, might adversely affect a local community dependent upon the forest for its livelihood. The Forest Service deemed properly regulated lumbering in a wilderness area compatible with the primitive area designation. Such a policy stance allowed for logging on about eighty thousand acres of the South Fork Primitive Area, one of three such areas that originally constituted what is now

In the sign image: **The GILA WILDERNESS AREA** FIRST NATIONAL FOREST AREA SO DESIGNATED EMBRACING THE MOGOLLON MOUNTAINS SEEN IN THE DISTANCE EXISTS IN LASTING TRIBUTE TO THE MEMORY OF **ALDO LEOPOLD** PIONEER IN WILDERNESS PRESERVATION WHO HERE INITIATED THE ESTABLISHMENT OF A NATIONAL WILDERNESS SYSTEM. THE WILDERNESS SOCIETY · FOREST SERVICE N. S. 2. 1. 1954

In 1954, thirty years after the Forest Service established the Gila Wilderness Area at the urging of Aldo Leopold, the Wilderness Society and the Forest Service erected this sign, which declared that Leopold's suggestion initiated the establishment of the National Wilderness System. (USDA Forest Service)

the Bob Marshall Wilderness in Montana. Greeley and his immediate successors handled each area on a case-by-case basis as the Forest Service continued fumbling toward more general long-term recreation and wilderness policies.[288]

For Forest Service landscape architect Arthur Carhart, Greeley's position on wilderness did not go far enough. In 1919, Carhart began working on the agency's first integrated recreational plan on the San Isabel National Forest, Colorado, where he developed the first architect-designed camp area at Squirrel Creek Canyon the following year. As part of his job, he assessed the area around Trapper's Lake in Colorado's White River National Forest for the construction of summer homes and a hotel in 1919. He insisted that all structures be kept at a distance of at least half a mile from the shoreline, with trails providing access to the lake. Though his proposal was eventually accepted, Carhart became frustrated with the Forest Service's level of support for recreation, which was hampered by a Congress uninterested in funding the agency's most basic recreational endeavors, and resigned in December 1922.[289]

While examining Trapper's Lake, Carhart had the opportunity to meet and exchange views with fellow Forest Service employee Aldo Leopold. In Leopold, Carhart found a philosophical soul mate. Leopold, who had become involved in wildlife protection and game management while creating a recreation policy at the Kaibab National Forest in northern Arizona in the 1910s, had grown concerned over the expansion of roads in the national forest and its impact on wildlife habitat and recreation. In 1921, he urged designating portions of the Gila National Forest in New Mexico as a wilderness area or national hunting ground; timber cutting and grazing would be allowed as long as no permanent roads accompanied them. Three years later, based on this recommendation, Chief Greeley designated 500,000 acres of the Gila as the nation's first wilderness area.[290]

Leopold's notion of wilderness expanded that of Carhart's. Both men wanted to keep roads out of wild areas to limit the encroachment of the modern world. But whereas Carhart's proposal protected a narrow strip around a lake, Leopold's protected a region. A national policy on wilderness remained in the distant future, but the idea of roadless wilderness areas had taken root.[291]

Robert Marshall and Recreation

In 1935, Aldo Leopold, who had written in favor of wilderness ten years earlier, joined with forester Robert Marshall and six others to form the Wilderness Society to advocate for the preservation of wilderness areas. In many ways, Leopold and Marshall picked up where naturalist John Muir left off. Whereas Muir stressed the value of nature and forests for the individual, Marshall applied the idea of nature's rejuvenating qualities to society and argued that the restorative powers of wilderness could help slow moral deterioration. By giving urban dwellers a respite from the problems of modern civilization, wilderness might turn people away from war and crime. The importance of forests in saving society also led Marshall to side with Gifford Pinchot in the 1930s in calling for the increased regulation of private timberlands and the expansion of public forests. Marshall put these themes into a pamphlet, *The Social Management of American Forests* (1930), and a book, *The People's Forests* (1933).

Marshall was the son of Louis Marshall, a prominent constitutional lawyer and social reformer who fought to establish the New York State Forest Preserve in the Adirondacks. The wealthy and prominent Jewish New York family spent its summers at the family home in the Adirondacks, where Bob developed a yearning to explore and a love of nature and hiking. A prodigious hiker, he regularly walked thirty to forty miles a day for relaxation, carrying only a bag of raisins and a hunk of cheese for nourishment. He purportedly once walked seventy miles in twenty-four hours to make a connection for a trip.[292]

After getting bachelor's and master's degrees in forestry, Marshall joined the Forest Service in 1925 and worked at the Northern Rocky Mountain Forest Experiment Station in Missoula, Montana. Three years later, he left and completed his doctorate at Johns Hopkins University before traveling to Alaska to explore and map the virtually unknown Brooks Range, above the Arctic Circle.[293] The Forest Service hired him for 1932 and 1933 to prepare the chapters on recreation for the *National Plan for American Forestry* (the Copeland Report), which offered a comprehensive plan for more intensive management of the national forests. In 1933, he became director of forestry in the Office of Indian Affairs, where he successfully argued for the retention

Bob Marshall, who shaped recreation policy while chief of the Division of Recreation and Lands, wrote, "Each mile on a river will take you farther from home than a hundred miles on any road." He recognized, however, that the agency must offer access to rivers *and* roads to provide a variety of recreational experiences. (Courtesy of The Wilderness Society)

of roadless areas on Indian reservations so that residents could withdraw from contact with whites.

From 1937 until his premature death at the age of thirty-eight two years later, Marshall served in the Forest Service as the chief of the new Division of Recreation and Lands. Prior to the division's establishment, the Forest Service had viewed recreation primarily as a matter of sanitation and fire prevention, and in the context of its rivalry with the National Park Service.[294] As division chief, Marshall urged greater protection for primitive areas in national forests. He was instrumental in the adoption of the "U regulations," many of which came out of his work on the Copeland Report a few years earlier. The U regulations of 1939 gave greater protection to wilderness areas by banning logging, road construction, summer homes, and even motorized vehicles from primitive and wilderness areas. With Forest Service leaders somewhat confused about how to define wilderness, let alone manage it, Marshall discussed the many purposes of recreation for the wide range of visitors to the national forests and proposed a recreational survey to help decision makers determine the preferred types of recreation and how much land to set aside for each.

Marshall defined recreation in simple terms for the purposes of the Copeland Report: "anything that is done directly for the pleasure or enrichment which it brings to life, in contrast with things that are done primarily to obtain the necessities of life." He proposed seven categories of recreational areas: "superlative areas," which would be "surpassing and stupendous in their beauty"; "primeval" or natural areas where old-growth stands would be preserved for scientific and therapeutic recreational use; "outing areas" just large enough to get away from highway noise; "roadside areas," or strips along major thoroughfares for scenic drives; organized and regulated campsite areas; residence areas for private homes, hotels, and resorts; and finally, "wilderness areas," places without permanent facilities, inhabitants, or mechanized transportation where a person could "spend a week or two of travel in them without crossing over his own tracks." (By comparison, Leopold thought a wilderness area should be at least large enough to accommodate a two-week-long horseback-riding excursion.)

Marshall recommended setting aside forty-five million acres, or nine percent of the total commercial forestland in the United States, for recreation. Of the forty-five million acres, he wanted ten million acres for wilderness areas. Another eleven million acres he recommended as outing areas with permitted economic development, such as logging.[295] In short, he proposed zoning the national forest for different uses.

Marshall's ideas reconciled several problems and disparate constituency groups. He wanted to allow grazing in western wilderness areas to continue because of the importance of livestock to the region's economy. An advocate of social welfare who was keen to the prejudices faced by minorities because of his own heritage, and a believer in the uplifting power of forests and recreation, he argued that the range of recreation categories would open the land to more people and pushed for placing recreational areas close to urban centers. People at the lower end of the economic scale, in the present and in the future, stood to benefit from forests waiting for them to visit. Although only a minority of people could afford a trip into his proposed wilderness areas, he believed their rights needed protecting, too. His support of the U regulations helped establish both easily accessible recreation areas and permanent wilderness areas.

Chief Gus Silcox understood what was really at stake. In 1935, he had publicly stated his support of multiple use. He recognized, however, that unless the Forest Service formalized its wilderness program, the agency stood to lose that land to the Interior Department. A survey of members of the Society of American Foresters in 1938 showed that by a margin of twenty to one, foresters favored legislation giving wilderness full protection.[296]

Marshall, Leopold, and Carhart successfully advocated on behalf of wilderness at the local level. The Forest Service responded to their proposals and ideas by administratively designating several wilderness areas across the country. Those areas, however, could be (and sometimes were) opened for logging or mining. By and large, with little industrial need for resources from the national forests before World War II, wildlife habitat and wilderness were preserved almost without effort. Historian Paul Hirt has observed that the lack of management proved to be good management for these resources. Yet despite the good-faith effort of the Forest Service, preservationists feared that future administrations might open up wilderness areas for incompatible economic development.[297]

The Postwar Recreation Boom

The escalating demand for timber from national forests during and after World War II exacerbated preservationists' fears. After the war, congressional and economic pressure to "get the cut out"—to meet the needs of the boom, keep the national economy strong, and preserve lumber communities—grew substantially in the late 1940s and 1950s. By then, multiple-use forestry was in full swing, with the forests under heavy use for recreational activities as varied as alpine skiing, rafting, and camping, along with mineral and timber extraction and the harvesting of nontimber products, such as mushrooms.

Recreation visits rose from 18 million in 1946 to 52.5 million ten years later. With the national forest road system providing greater access, recreational use increased from 27 million to 178 million recreation visitor-days per year between 1950 and 1971.[298] The amount of land used for recreation, however, did not increase to meet that demand. Primitive, wild, and wilderness areas remained at the 14-million-acre level (just over seven percent of the national forests) from 1935 to 1958. Often, diverse activities took place within a few miles of one another. Despite Forest Service efforts to educate the public through books and television shows, the general public made little distinction between national forests and national parks and questioned the lack of "preservation" on the national forests. Numbered among the agency's supporters were members of the wilderness movement and recreation groups who wanted to move large tracts of land from the commodity extraction category to a purely recreation one.[299]

In 1957, a bill calling for the protection of wilderness areas was introduced in Congress. The Forest Service wanted to avoid a wilderness bill that would reduce its power to decide what was best for lands under its jurisdiction. It countered the wilderness bill with "Operation Outdoors" in 1957, a five-year program for the development of recreational facilities on the national forests. Like the National Park Service's ten-year recreation program, "Mission 66," submitted to Congress the year before, the Forest Service plan attempted to establish long-term objectives and cost estimates to win administration and congressional support for substantial funding increases, as well as compete with the National Park Service.[300]

The Forest Service also hoped the plan would cultivate, if not capture, recreation and wildlife constituencies, who were increasingly frustrated with the Forest Service. Those constituencies had other options for their recreation dollars and time besides national forests, including land administered by the National Park Service and Bureau of Land Management (created in 1946 when the General Land Office and the Grazing Service merged); Tennessee Valley Authority, Army Corps of Engineers, and Bureau of Reclamation projects; wildlife refuges; and state and private lands. They were not reassured when President Dwight Eisenhower's administration forced the Forest Service to shelve the wildlife portion of Operation Outdoors, though the agency released it when a new presidential administration supportive of wilderness programs came to power in 1961.[301]

As promoted by the Eisenhower administration, Operation Outdoors focused much-needed attention on recreation problems, and the federal government responded positively, more than doubling funding from $4 million to more than $9 million for the 1958 fiscal year and steadily increasing funds throughout the remainder of Eisenhower's term. The agency not only had money for facilities construction for the first time since the New Deal, but it had the money for a full-time recreation staff for the first time in its history. The new staff augmented the agency's limited professional expertise beyond forest engineering and timber management. At this point, the Forest Service began to function as a true multiple-use agency, though one still focused on timber as the land's primary use.

Even the Multiple Use–Sustained Yield Act of 1960, initially opposed by the timber industry when first drafted in 1956, did not change the status quo. There was nothing in the law stopping develop-

ment, which pleased the timber industry. In fact, industry supported the bill after Chief Richard McArdle, who succeeded Lyle Watts in 1952, had assured them it would not undermine or alter the watershed and timber aspects in the 1897 Organic Act.

The legal limit for game birds in Texas in 1916 and in many other states was not sustainable. The Forest Service's State and Private Forestry branch has worked with state game wardens and state legislatures to reform hunting laws since 1908. (USDA Forest Service)

Early Wildlife Management

President Benjamin Harrison created Yellowstone Timberland Reserve, the first forest reserve, in part to provide protected habitat for the few remaining bison and elk. By then, however, deer had been hunted nearly to extinction in the East, and bighorn sheep, moose, and woodland caribou remained only in remnant herds in their native habitats. Aquatic species were fairing little better. The decline in freshwater and marine fish populations had become noticeable by 1900. To address these problems, the Department of Agriculture's Bureau of Biological Survey (later the U.S. Fish and Wildlife Service in the Interior Department) and state agencies undertook game preservation and propagation efforts.[302]

When the Forest Service was created in 1905, Congress and the courts had already defined its administrative responsibilities for wildlife management. Wildlife on national forests remained under the jurisdiction of the state, and the Forest Service managed the land. To avoid conflict with individual states, the Forest Service in 1906 authorized rangers to cooperate with state game laws and state officials and simultaneously serve as deputy state game wardens. In 1911, the federal government allowed some states to establish wildlife refuges on national forest land. By 1915, most states had enacted fish and game protection laws and established administrative agencies to implement and enforce them. In reality, though, few localities outside the Northeast

Unlike game animals, predators were not protected from overhunting by law, and many were hunted to near-extinction. In many states, bounties for hides made them all the more attractive to hunters and Forest Service personnel charged with eradicating them. These predators were caught on the Shoshone National Forest in Wyoming, 1914. (USDA Forest Service)

and progressive pockets in the Southwest cared to spend money on enforcement until the late 1930s.[303]

At the beginning of the twentieth century, wildlife management was hampered by the limits of biology and its subdiscipline ecology.[304] Early ecologists were particularly concerned with population dynamics and the mechanism of population control. Attempts to control pests and predators led ecologists to conclude that detailed study of the life histories of all the species involved was required. Some long-term studies to determine the complex biological relationships between smaller predator and prey and between plants and animals were under way when the Forest Service was established, but ecological studies of larger predators, such as wolves, and their prey presented technical problems that discouraged researchers until the 1960s.[305]

As with many aspects of land management, the Forest Service treated wildlife management as a matter of economics—a way to advance the nation's economy. Certain animals were commodities valued for their hides or meat; predators and animals with little or no commercial value were to be ignored, limited, or in some cases eliminated. State game and fish commissions, many formed in the late 1800s, shared this perception and took measures to protect commercial fisheries and game but not noncommercial animals. In many parts of the country, local governments offered a bounty for predators. There was little understanding of the ecological role each species played in nature.

Competing constituencies tended to influence the focus of federal scientific work. Biologists with the Biological Survey told farmers that wolves were beneficial because they ate rodents, while at the same time the bureau actively supported a predator-eradication program to kill wolves for ranchers, who paid fees to have the bureau carry out predator hunts in the 1920s.

The fees eventually accounted for one-quarter of the budget, and the ranchers' viewpoint came to dominate the bureau's position on wolves.[306]

The Forest Service was in a similar situation. Ranchers and hunters, who paid fees to the Forest Service and expected something in return for their money, also wanted predators brought under control. Consequently, although the agency's game management position emphasized propagation (mostly through artificial restocking) to produce deer, elk, and moose for hunting, it gave equal emphasis to eradicating cougars, wolves, and coyotes to protect game and livestock.

Federal efforts in game management initially met with mixed results. The U.S. Fish Commission, which was established in 1871, carried out fish propagation programs that restored shad and salmon stocks.[307] Efforts on the Columbia River largely failed, however, when an estimated ninety percent of the eighty million to ninety million salmon fry wound up in irrigation ditches. A breeding program for bison, whose numbers had dwindled to 970 in 1905 from perhaps a billion in the 1830s, did succeed, in large part because bison herds roamed in protected game preserves; a small number were propagated on the Wichita National Forest in Oklahoma, now a national wildlife refuge. The Forest Service's facilities on the Pisgah National Forest for studying and rearing deer, including a laboratory and a specimen collection of flora and fauna, were judged by one Biological Survey scientist as valuable sources of deer restocking equal to any he had encountered.[308]

Crisis on the Kaibab

The most famous example of the shortcomings in the agency's game management policy began in the late 1800s, when mule deer populations around the Grand Canyon's Northern

Early fish propagation efforts were rudimentary at best. This stream on the Black Hills National Forest was restocked sometime around 1935 with fish transported by horse and buggy in what appear to be milk containers. (USDA Forest Service)

Rim declined dramatically because of excessive hunting. When President Theodore Roosevelt created the Kaibab Game Preserve in 1906, some three thousand to four thousand mule deer shared year-round range with thousands of cattle and sheep. To grow the deer herd, the government banished livestock and banned game hunting but encouraged government employees to kill predators. Hunters killed wolves, coyotes, mountain lions, and bobcats so efficiently that by the 1920s, some sources estimated the size of the Kaibab deer herd at one hundred thousand. Because the preserve was surrounded by desert, the deer could not migrate to other areas. Their forage depleted, they began to starve.

After studying the feeding habits of the deer, Ranger Benjamin Swapp unsuccessfully recommended opening the game preserve to hunting to reduce the deer population in 1920. Habitat conditions continued to deteriorate. In 1923, nearly 20,000 deer starved to death. The following year, Secretary of Agriculture Henry C. Wallace sent a committee of wildlife and grazing interests to investigate. They, too, recommended reducing the herd. Because the preferred method was to transplant—not kill—the animals, the Arizona Game and Fish Department, with the permission of the Forest Service, allowed a "deer drive" in an attempt to move a large number of them to a new location. The spectacle of men on foot and horseback clanging tin cans, cowbells, and other noisemakers was "the most interesting failure" one ranger had ever seen.[309]

Left with little choice, the Forest Service sent in rangers to shoot the deer, triggering a conflict over who had jurisdiction over wildlife management on the national forest. The state opposed the hunt and arrested three hunters. In the resulting lawsuit, coincidentally titled *Hunt v. United States*, the U.S. Supreme Court ruled in favor of the Forest Service in 1928, noting that the federal government had a right to protect its property if excess game populations endangered it. This established a crucial precedent for federal control of wildlife on federal lands.

Twelve years later, the courts upheld the Forest Service position again in *United States v. Chalk* on the same grounds as in *Hunt v. United States*—the right of the federal government to protect its property.[310] By this time, though, the policy permitting hunts had raised such an outcry from state game commissioners around the country that the Forest Service revoked its controversial Regulation G-20A, which authorized hunts on national forests, and passed a new regulation, W-2. It made the states primarily responsible for the protection and use of wildlife and fish on the national forests; the Forest Service would be primarily responsible for managing these species' habitat.[311] The events generated extensive coverage in conservation, outdoors, and scientific publications and demonstrated the need for a more complex approach to wildlife protection.

Aldo Leopold and Wildlife Management

In Forest Service and environmental history, Aldo Leopold's ideas mark the intersection of wilderness, wildlife management, and recreation policies. Leopold was also one of the first in the Forest Service to link forestry and ecology. Though he published most of his work on ecology

and issued his call for a land ethic after he had left the Forest Service, his experiences while working for the agency informed and shaped his thinking. Given his influence on recreation, wilderness, wildlife management, and ecology, Leopold's impact on public land management is immeasurable.

Leopold joined the Forest Service after graduating from Yale's forestry school in 1909, and by 1912 he was supervisor of the Carson National Forest in northern New Mexico. In contrast to the Forest Service's focus on single-species propagation, Leopold had begun moving toward an ecological approach. In one of his first publications, a 1913 letter to his fellow officers that appeared in the Carson National Forest newsletter, he laid out virtually all the purposes of the national forests—"timber, water, forage, farm, recreative, game, fish, and esthetic resources"—that nearly a half-century later would be codified in the Multiple Use–Sustained Yield Act of 1960. But for Leopold, the measure of successful management would be "the effect on the forest"—the environmental impact of activity on the forest. Successful management went beyond adhering to official policy, following procedures, and meeting timber quotas or counting dead predators. Leopold had begun to view the forest as an ecosystem.[312]

In 1915, the Forest Service assigned Leopold to develop a recreation policy for the Kaibab and Tusayan national forests, which surrounded the Grand Canyon. The new job included coordinating

Aldo Leopold was photographed with a friend's dog on the Apache National Forest about a year after he joined the Forest Service in 1909. While leading a reconnaissance crew on the Apache National Forest around this time, he had his encounter with a wolf that changed his attitude toward humans' relationship with the environment.
(USDA Forest Service)

the nascent fish and game program. Leopold took advantage of the opportunity and began his game management studies and wildlife surveys while working on plans for a game refuge. His *Game and Fish Handbook* for Forest Service rangers and officers, the first of its kind, brought him attention from the national office. His involvement with game preservation associations in the Southwest led to his emergence as a leader of the national game conservation movement.[313]

After fifteen years in the Southwest, Leopold became the associate director of the Forest Products Laboratory in Madison, Wisconsin, in 1924, just as the Kaibab disaster was unfolding. His new position gave him the luxury of examining the problem from a distance and without having to justify and defend policy. By this time, he and a small group of zoologists and foresters had concluded that effective game management required more than just control of hunting or eliminating predators. It meant managing the species' environments to maintain proper habitat conditions.[314]

Leopold left the Forest Service in 1928 to conduct wildlife game surveys throughout the country. A few years later, he combined his survey statistics with the theories of British ecologist Charles Elton, one of the prominent figures in the development of animal ecology, to develop game management into a rigorous applied science.[315] Elton, in his 1927 book, *Animal Ecology*, presented principles for integrating population and community ecology—niches, the food chain and food cycle, food size, the "pyramid of numbers," and others—that underlay the new science of game management. The principles, when combined with directed research and quantitative studies, supported the notion of the "balance of nature." Leopold and Elton met in 1931 and quickly struck up a cooperative friendship. Leopold's work provided Elton with the field research he needed, and Elton's ideas contributed to Leopold's development of game management as a science.[316]

Applying the principles of forestry, such as sustained yield, to game management gave it immediate professional standing. Leopold's position—that farm and forest game could, and should, be raised like crops for economic gain—led many in game management to ignore ecological ideas then emerging, ideas that Leopold himself supported.[317] In 1933, he accepted an appointment to the newly created chair in game management (later wildlife management) at the University of Wisconsin, a position he held until his death in 1948. He trained a generation of leaders in game and wildlife management and disseminated his ideas about ecology. In their later work with hunters and other sportsmen, his students diffused these ideas to the public.

The same year he began teaching, Leopold published *Game Management*, the text that helped define the profession and is still in use today. In *Game Management*, he renounced his previous support of killing predators but also criticized those "students of natural history," or preservationists, who supported no predator control at all: "Both extremes are biologically unsound and in many cases economically impossible. The real question is one of determining and practicing such kind and degree of control as comes nearest serving the interests of all four groups [agriculturists, game managers and sportsmen, preservationists, and the fur industry] in the long run."[318] His

use of "in the long run" showed game management's conservation roots and the science's underlying mission of sustainability.

Even as he prepared the textbook, Leopold was moving away from its utilitarian approach and coming to think of all wildlife as part of a natural system. Initially, his game management studies revealed the importance of predators as a check on game populations. He also drew on his own experience on the Kaibab National Forest, which upon later reflection led him to rethink his position. While enforcing the Forest Service's antipredator policy in the Southwest, Leopold had a dramatic encounter with a wolf that changed the direction of his life's work. After shooting and wounding a mother wolf, he found her in time to witness "a fierce green fire dying in her eyes." Reflecting on the incident in the essay "Thinking Like a Mountain," written in 1944, he cited it as the beginning of his philosophical shift from thinking of wilderness in commercial terms to understanding it as a place where the complex ecological interdependence of humans and animals played out in the American landscape.[319]

What moved his ideas along the continuum was his time at "the Shack." In 1935, Leopold had purchased a worn-out, 120-acre farm in the "sand counties" area along the Wisconsin River and set about rehabilitating the land by restoring its ecological balance. He and his family planted trees and converted a chicken coop, the only standing building on the property, into a cabin the family called the Shack. The rehabilitation efforts over the next dozen years, all noted in

Aldo Leopold's "Shack" about five years after he and his family began restoring the structure and the exhausted land around it. Leopold's time at the Shack had an enormous impact on his understanding of ecology. (Aldo Leopold Foundation)

numerous journals that he later drew upon for writing books and articles, taught him "a love and respect for the land community and its ecological functioning."[320] Spending time on the farm, which included a marsh, gave him an opportunity to study the interaction of predator and prey and observe the daily and seasonal changes, as well as to restore breeding habitat for several wildlife species.

In the wake of the Kaibab disaster, the Forest Service launched studies on predator control, game population dynamics, and the relationship between habitat degradation and declining species. The Kaibab disaster also led to closer cooperation between the Forest Service and other agencies in their efforts to understand better the relationship between game populations and livestock and other forestry issues as they affected the public welfare. But none of this work had an ecological basis. Agencies and preservation groups continued working to save a single species at a time, with emphasis on predator control. Land management remained anthropocentric, focused on what was best for humans and how to get what people wanted from the land.

In the mid-1930s, as ecological ideas gained acceptance in wildlife science circles, the word game fell out of favor, to be replaced by "wildlife." The change reflected the shift toward understanding the importance of nongame animals and their habitat. Concern over the long drought during the 1930s led to legislative action on behalf of wetlands and dwindling bird populations, with several federal laws aimed at promoting wetland restoration and wildlife research.[321] In 1937, Congress passed its most far-reaching measure yet, the Federal Aid in Wildlife Restoration Act, better known as Pittman-Robertson. The act authorized a ten percent federal tax on sporting arms and ammunition, with the proceeds to be given to the states for wildlife research and land acquisition, development, and maintenance.

Division of Wildlife Management

The Pittman-Robertson Act required the states to employ trained personnel for their wildlife programs and thus fostered the professionalization of wildlife management. In 1936, Aldo Leopold had joined with wildlife biologists and others working for the protection of wildlife populations and their habitats to form the Wildlife Society, the first professional organization in the discipline. In addition, the Forest Service established its Division of Wildlife Management and assigned sixty-one people to wildlife work that year. They carried out cooperative agreements with state wildlife commissions to jointly manage designated game areas and discourage poaching. Though timber management received the highest priority throughout this era, the agency made great strides in wildlife management, such as the establishment of a game preserve in the Pisgah National Forest to serve as a model and testing ground for deer management.

In the months before World War II, the Forest Service began shifting its funds and focus toward timber needs as the nation prepared for war. During the war, wildlife biologist Lloyd Swift, appointed the Wildlife Division's second director in 1944, nonetheless anticipated a time when wildlife management would encompass more than simply game management. Swift issued a memo calling for "broadening the wildlife approach so that more attention is given to the betterment of fish, upland game birds, fur bearers, vanishing species, etc.…Postwar and subsequent work should…consider wildlife for its own values and particularly the value it has in providing public recreation." Continually high demands for timber, though, meant that more than three decades would pass before the federal government fully embraced his views.[322] In fact, Swift saw a significant drop in funding for wildlife management over the course of his twenty-year tenure, and the declines were not always the fault of the Forest Service. From 1946 to 1949, for example, Congress refused to budget any money for wildlife work. Salaries for Swift and several regional specialists had to be paid out of discretionary funds. Some money came from the funds provided by the Pittman-Robertson Act, but when states realized the Forest Service was doing little for wildlife management, they withheld funding, too.[323]

In the postwar era, however, the Forest Service continued to approach wildlife as game production, secondary to timber, even as the number of visits by hunters and anglers jumped substantially, as did membership in wilderness organizations. Interest in outdoor recreation and wildlife differed from that of the interwar years in one significant way. The public increasingly saw animals less as game and more as indicators of environmental health. Ironically, Americans' rising standard of living often depended upon practices that destroyed the wildlife they wanted to protect, such as the use of chemical fertilizers and pesticides in agriculture and the growth of suburban housing developments in previously rural areas.[324]

Continued emphasis on timber and clearcutting kept wildlife management a low priority through the 1950s until governmental biologists in both the Forest Service and other agencies, who had long supported the maxim that "what's good for timber is good for wildlife," called clearcutting into question. Wilderness advocates who looked at Forest Service timber management plans grew alarmed by the rising projections of timber harvest levels and the increasing reliance upon clearcutting to achieve them, and they began calling for a national wilderness preservation system.

Meanwhile, the wilderness and wildlife movements had become intertwined. Wildlife conservation groups supported wilderness preservation measures, hoping both would benefit in the long run. Wildlife management, which had only received intermittent attention from the agency, benefited from public interest in wilderness protection and the recreation boom. Wilderness and wildlife advocates turned to Congress for legislative help in the late 1950s, but it was a decade before Congress responded with the legislation they wanted. The 1950s ended with the foundation for change having been laid.

Chapter Six

THE POSTWAR PERIOD AND THE RISE OF THE ENVIRONMENTAL MOVEMENT

A Pacific Northwest shelterwood cut in the 1960s. (Jim Hughes – USDA Forest Service)

FOLLOWING WORLD WAR II, THE FOREST SERVICE BEGAN RECEIVING MIXED MESSAGES FROM THE PUBLIC ABOUT LAND USE. ON THE ONE HAND, LUMBERMEN HAD DRAINED MOST PRIVATE INDUSTRIAL LANDS OF COMMERCIAL TIMBER AND HAD TURNED TO THE NATIONAL FORESTS TO SUPPLY THE BOOMING HOUSING AND CONSUMER MARKETS OF THE COLD WAR ERA. THE TIMBER INDUSTRY EXERTED CONSTANT PRESSURE ON THE FOREST SERVICE TO RAISE ANNUAL TIMBER HARVEST LIMITS ON NATIONAL FORESTS. FOR ITS PART, THE AGENCY OFTEN MEASURED SUCCESS FOR DISTRICT RANGERS AND REGIONAL FORESTERS BY THE AMOUNT OF BOARD FEET EXTRACTED, NOT BY THEIR FORESTS' QUALITY OF HABITAT, VIABILITY OF WILDLIFE POPULATIONS, OR NUMBER AND QUALITY OF RECREATION EXPERIENCES. THE FOREST SERVICE REMAINED LARGELY CONCERNED WITH TIMBER MANAGEMENT AND SOON FOUND ITSELF IN AN ENVIRONMENTAL COLD WAR, WITH MUCH OF THE SKIRMISHING OCCURRING IN CONGRESS AND THE COURTROOM.

What was happening in the national forests led some preservationists and politicians to advocate limiting or overseeing Forest Service activity. The efforts of Bob Marshall, Arthur Carhart, and Aldo Leopold to establish wilderness and recreation policies also led to new legislation setting aside wilderness areas. The publication of a popular science book that discussed the harmful effects of pesticides touched off even greater efforts to protect the environment. It also launched a reexamination of the agency's land management practices that resulted in new laws that completely changed the way the Forest Service operated.

Zoning the Land

Private timber stocks had dwindled during World War II, and the Forest Service obliged private industry by opening the national forests to meet demand. For Forest Service district and regional foresters, increased cutting of federal timber meant greater opportunities for promotion and advancement, which reinforced a stronger belief in clearcutting. But it was more than simply what was good for the individual forester. His work was part of the larger Cold War struggle. The demands of constantly being on a war footing and building up a bulwark against communism strongly influenced natural resource policies.

President Harry Truman appointed the President's Materials Policy Commission in January 1951 to assess the situation and outlook. Its report, Resources for Freedom, came out the following year and called for developing natural resources quickly to defeat communism: victory required the elimination of poverty and social inequalities by spreading prosperity. A house for any family

The postwar home construction boom consumed large quantities of timber products. The Forest Products Laboratory developed the truss frame system to speed construction and reduce waste and, in turn, lessen demand for lumber.
(USDA Forest Service)

that wanted one was evidence of material wealth. Since housing construction consumed large quantities of timber for framing, paneling, plywood, fiberboard, particleboard, and furniture, it became a national duty and moral imperative to maximize timber production. The commission called for the construction of some six thousand miles of timber access roads in the next five years. Meeting quotas meant more than personal prosperity; it meant victory for democracy.[325] Cold Warriors at the Pentagon talked throw-weights and missile counts; foresters talked timber volume and board feet.

By the time Lyle Watts stepped down as chief in 1952, the pattern of annual increases in timber production from federal lands was firmly established. To meet rising demand, the Forest Service would produce timber more efficiently by increasing the use of even-age management—clearcutting followed by replanting. The increased federal funding that supported it reinforced its reputation as the best silviculture method available. Timber projection models indicated that more even-aged stands would be needed to meet future demand.

The Forest Service got caught up in the circular logic, and every decision became a rationalization that supported more clearcutting. The Eisenhower administration quickly took the original meaning of sustained yield to its logical conclusion to mean increasing timber harvesting to reach the maximum allowable cut—the point at which timber cut was equaled by replacement. Clearcutting as practiced by foresters promoted efficiency, met timber demands, and provided economic benefits for private industry and public coffers alike; mature forests whose growth was slowing could be replaced with younger, faster-growing trees, which in turn meant shorter cutting rotations. Some species could be harvested every 60 to 80 years, for example, instead of 150 or more.[326]

Budgetary increases for timber management in the 1950s further fueled the drive to get the cut out. Some incentives had been on the books for some time. A 1913 law automatically returned ten percent of gross timber receipts to the agency for road construction. Although timber sales meant jobs and income for communities, higher sales receipts also meant the agency retained more money through its trust funds (as set up by the Knutson-Vandenberg Act of 1930 and other similar laws) and ensured that western congressional representatives would support the Forest Service at budget time. The Knutson-Vandenberg Act funneled most of the remaining timber receipts (after counties were given their share) to the agency for reforestation or timber stand improvement, and thus offered a powerful incentive for district and regional foresters to sell more timber.[327]

New technologies, such as chainsaws and war surplus trucks and tractors, combined with a massive road construction effort to make more timber accessible, opened up the national forests like never before. Lumber production on national forests quickly rose from 4.5 billion board feet in fiscal year 1952 to 7 billion five years later, with a corresponding increase in revenue from the sales. Timber sales clearly demonstrated that the Forest Service produced much-needed commodities and contributed to the Cold War effort in a very tangible way. In the pro-business atmosphere of the 1950s, the Forest Service even seemed capable of making good on Gifford Pinchot's forty-year-old promise to pay for itself.[328]

Under Chief McArdle, appropriations in the 1950s for timber access roads and forest highways rose. The change in funding for road construction was a consequence of the emphasis on timber management. Previously, a timber sale operator built a road and controlled access to it, thereby maintaining control over sales in the area. Publicly funded roads meant public control. Congress, which had been funding construction of the federal highway system since 1944 and had embraced Eisenhower's interstate highway system, agreed with McArdle's suggestion that the public pay for and control the roads, and construction of new roads in the national forests with taxpayer money soon began. The decision put the Forest Service in the road-building business. The national forest road system nearly doubled from 107,000 miles in 1950 to nearly 200,000 in 1970. In that same period, timber sales increased from 3.4 billion to 10.6 billion board feet per year.[329]

New logging roads, like this one on the Boise in 1955, permitted the Forest Service to increase timber sales, but when the public saw logging operations up close (facing page), opposition to clear-cutting began to grow. The logging operation was in northern California. (USDA Forest Service)

Meanwhile, the Forest Service sought to reassure the public that lumber companies carried out their operations on public land under Forest Service regulations. Clearcuts, the agency said, fit into multiple-use plans by providing openings that benefited big game like deer and elk, and forest road expansion improved access for the general public. Those same roads brought the public face to face with how the Forest Service handled the land, however, and it was not long before some began voicing disapproval.

Although proud of its timber management program, the Forest Service was sensitive to criticism and adopted practices to hide deforested land from public view, such as the use of "beauty strips," the debate over which went back to the early days of automobiles in the national forests. In 1924, Forest Service leaders first considered leaving buffers of uncut trees, varying from 100 to 200 feet wide, along roads in sale areas to mask logging activities. As motorists drove through national forests, they had encountered logging and complained. J. F. Preston in the forest management staff argued against the buffers and instead urged carrying out logging in full view to demonstrate what the agency was trying to do. Herbert Smith, who handled public relations and information, pointed out that Bernhard Fernow and the Cornell forestry program met with disaster in the early 1900s when he showed the public what he was doing in the Adirondacks. Chief Greeley argued there was nothing to hide, and that logging was a positive aspect of management that showed the agency's dedication to using the land. In fact, the Forest Service reserved the right to log areas as part of fulfilling its mission. T. W. Norcross, head of engineering, said it did not matter what the agency thought; the public favored recreation, and that should dictate the Forest Service's decision. Greeley tabled the issue.[330]

Nevertheless, beauty strips became more common as both timber harvesting and nontimber use of the national forests increased over the next three decades. Meanwhile, the Forest Service kept up its attacks on private industry for clearcutting in the Northwest and continued pushing into the 1950s for regulatory powers to stop it. By then, the agency, which was coming under increasingly sharp criticism for its recreation and wildlife policies, wanted to obscure the fact that it had embraced clearcutting, and that it had, in effect, completely reversed its position on timber management.[331]

Besides masking clearcuts from casual observers and critics alike, beauty strips were the first step toward developing a way to manage simultaneous access to multiple resources through the use of zones. In the 1950s, the agency embraced the "zoning principle" for handling the land. "Zoning" the land would place different activities on the parcels of land most suitable for that activity and thus minimize conflict between land users. Although multiple use did not necessarily mean that all of the important uses would occur on the same acre, declared Assistant Chief Forester Edward P. Cliff in 1953, timber production, a national priority, did receive priority

over other uses. The agency integrated recreation, grazing, and wildlife "as fully as possible without undue interference with the dominant use," but with the intent of achieving "the maximum *continuous* yield of beneficial products and services from the national forests."[332] Zoning combined the principles of multiple use and sustained yield and allowed the Forest Service to continue emphasizing timber production.

Multiple Use–Sustained Yield Act of 1960

By the late 1950s, as timber cutting reached areas frequented by their members, the Sierra Club and the Wilderness Society began voicing opposition to the emphasis on timber. The fight between the agency and the timber industry over regulation of private timberlands, which had ended by 1953, put the two former opponents on the same side of the argument. When the National Lumber Manufacturers Association adopted a policy statement supporting multiple use that echoed Ed Cliff's remarks, it did little to alter public perception about either organization.[333]

The Forest Service was struggling to retain authority over categorizing land on all national forests. Chief McArdle hoped that the agency's support of multiple-use legislation could better enable it to withstand timber industry pressure to give priority to timber and water (the only uses specifically mentioned in the Organic Act of 1897) at the expense of recreation and wildlife and fish. But McArdle also expected the Multiple Use–Sustained Yield Act of 1960 (MUSY) to shield Forest Service administrative control over wilderness lands within the national forests from congressional meddling.

MUSY expanded the 1897 Organic Act, declaring, "The national forests are established and shall be administered for outdoor recreation, range, timber, watershed, and wildlife and fish purposes."[334] By not defining those five "purposes," it allowed the agency some latitude in land management decisions. Development could continue, which pleased the timber industry. The act codified many long-standing practices. For the first time, the Forest Service had a specific congressional directive that established priorities for resource use and stipulated that economic return was not in all cases to be the deciding factor. After fifty-five years of creating policies for specific uses or resources on an ad hoc basis, the Forest Service now managed the uses under one unifying statement. The agency had long practiced multiple use; now it was the law of the land.[335]

As prescribed by MUSY, the Forest Service retained control over wilderness areas within the national forests. Notably missing from the law, however, was the mention of any restrictions on the *uses* of wilderness or other lands. Nor did Congress provide sufficient funding to change priorities. Consequently, the Forest Service took several steps to appease preservation and recreation groups. Timber plans adjusted logging activities to make them less disturbing aesthetically. The agency hired landscape architects to make clearcuts conform to the contours of the land instead of following a rectangular pattern, and to lay out beauty strips to help disguise activities.[336]

Preservationists soon realized that rising timber harvest levels, especially in the Pacific Northwest, affected the quality of nontimber resources. In the eyes of preservationists, instead of restraining the timber industry, the Forest Service had become part of the problem. The public's trust in land management agencies in general, eroding since the 1940s because of a series of decisions that affected recreation in national parks and forests, declined even faster when the Bureau of Reclamation tried unsuccessfully to construct dams in Dinosaur National Monument and Echo Park in the 1950s. Preservationist organizations learned to fight on the government's terms. They mastered the scientific jargon and taught themselves how the bureaucracy worked to undercut agency arguments supporting the dam, all the while marshaling public opinion on unprecedented levels. The victory energized the growing preservation movement as well as the public commitment to the preservationist mandate of the National Park Service Act.[337]

Though not involved in the Dinosaur Monument controversy, the Forest Service suffered in the fallout. Stopping one government scientific agency convinced the preservationists that they could take on others, and that they could use other government agencies as leverage. The National Park Service—the beneficiary of national attention on the Dinosaur controversy—once again became an alternative agency for protecting public forests, though worried preservationists noted that even the Park Service had done little to stop the dam project.

Congress, however, still had little interest in funding activities that generated little revenue and continued to allocate most of the Forest Service budget to timber and timber-related activities. From 1954 to 1970, the Forest Service received sixty-six percent of budget increases requested for timber sale administration, but only twenty percent of its requested increases for recreation and wildlife, seventeen percent for reforestation, and fifteen for soil and water management.[338] It was common for logging to be planned even in recreation areas. Even as it passed legislation requiring the Forest Service to practice multiple-use management, Congress failed to provide the funding necessary to ensure that the uses were in balance. Multiple-use plans based on the Washington office's expectations of meeting timber targets flowed down from the regional offices to the ranger districts. The days of independent, decentralized management were over. From the district ranger level on up, forest planners had to acquiesce to headquarters or risk their careers. Incentives based on timber harvest targets remained; no incentives came from Congress or Forest Service leaders for soil, water, or wildlife conservation. In reality, MUSY simply meant the Forest Service had to consider the effects of logging on other resources when making timber plans. Timber remained king.[339]

This is not to say that forest managers were unresponsive, or that all national forests focused solely on timber production. The Forest Service maintained or improved wildlife and fisheries habitats, particularly for game and targeted species, such as the condor and osprey. In 1960, the Forest Service took charge of nineteen national grasslands areas comprising some 3.8 million acres, mostly in the Great Plains, and continued restoring them so that they could be managed for multiple uses as defined by the Multiple Use–Sustained Yield Act. Rehabilitation work on eastern national forests and western rangelands brought positive results.

Critics, however, saw new roads leading to more clearcuts. The Forest Service called it multiple-use management because the roads opened up access to recreational activities, improved forage for game, made fighting fire and insects easier, provided wood for domestic uses, and created jobs for timber workers, but its promotion of multiple-use management through press releases and books and movies avoided talk of clearcutting. As far as the public knew, the agency still opposed the technique. Not informing the public of this shift before extending the practice to eastern hardwoods—and placing clearcutting operations closer to larger population centers without adequate explanation of the rationale behind it—set the stage for conflict.[340]

Communication within the agency was not much better. The Forest Service's hierarchical administrative structure and legacy of decentralized management left national forest managers to deal with local problems with little knowledge or understanding of what was occurring across districts, regions, or ecosystems, or in other scientific fields.[341] Integrated management could not yet begin without that information.

The decision not to fully inform the public about its practice of clearcutting, combined with a reluctance to change operations, created a problem of perception. Because internal change moved incrementally—improvements in the quality of managing multiple uses, in science and management skills, in restoration efforts, in responses to local demands and environmental concerns—in everything but timber cutting, the public perceived no change at all.

In the meantime, several warning signs indicated that segments of the public were no longer supportive of the continued emphasis on timber at the expense of other forest uses. The Forest Service was losing money on road construction and maintenance costs as it entered remote and rugged areas to gain access to timber. To justify these below-cost timber sales, the Forest Service said the roads provided access for wilderness recreation and firefighting and pest control; critics said the roads "actually marred scenery, damaged wildlife habitat, and caused severe erosion."[342]

The Wilderness Act

The continued threat to wilderness and the Forest Service's administrative control of wilderness areas made Howard Zahniser of the Wilderness Society and other wilderness advocates increasingly uneasy. Through the 1950s, the society pressed for federal legislation to give Congress the authority to set aside potential wilderness areas the Forest Service could not develop. Democratic Senator Hubert Humphrey of Minnesota and others wanted to ensure that land set aside for wilderness protection could not later be reclassified and opened for development, as had been happening with increased frequency in the national forests. The Eisenhower administration's agreement to withhold fifty-three thousand acres from the proposed Three Sisters wilderness area on the Willamette National Forest in Oregon in 1957 had been the latest example of the federal government's indifference to local public opinion.[343]

Senator Humphrey assembled bipartisan support for the bill, but it was opposed by the Forest Service, faced significant obstacles in Congress, and had little support in the Republican White House. Congressional opponents pointed out that the wilderness concept itself violated multiple-

Within the image: BURNED WATERSHED · ...ODED WATERSHED · SMALL CHECK DAMS

use principles because it prohibited extractive activities in favor of recreation for a handful of people. Several opponents of the bill invoked Gifford Pinchot and other early progressive conservationists and their ideas of developing natural resources to preserve the land. A lobbyist for water development interests stated at one hearing, "It is truly refreshing to note that neither Theodore Roosevelt nor Gifford Pinchot ever for a moment advanced the theory that conservation meant to set aside vast areas of a watershed where all of the natural resources should be bottled up and denied to the public except as something to look at." The latter-day conservationists defined conservation as "wise use," or planned development, and rejected any definition that combined rational use and deliberate nonuse. As Representative Wayne Aspinall (D-CO), who used his House Committee on Interior and Insular Affairs to repeatedly block wilderness bills, declared in 1963, "Traditionally, conservation and wise use have been synonymous...I do not know when, where, or how the purist preservation group assumed the mantle of the conservationist."[344]

The appropriation of Pinchot by resource industry lobbyists helped turn traditional conservation groups away from Pinchot and the term *conservation*. Pinchot, dead for nearly twenty years, came to personify senseless federal land development in the eyes of the growing environmental movement. Attracted to the writings of Aldo Leopold, John Muir, Bob Marshall, and others, some wilderness advocates increasingly used the word *environmental* and called for "ecocentric," or ecology-focused, rather than anthropocentric land management.[345] The participants on both sides were not unified in their thinking, but they were willing to paper over the differences of opinion within their own movements as long as the battle was joined against the opposition.

Undeterred—and untrusting of the Forest Service—Zahniser and his supporters continued pressing Congress for statutory protection. Using the Forest Service's zoning principle, they hoped to place wilderness in its own zones on the national forests. Senator Humphrey sponsored legislation drafted by Bob Wolf, a forester working for the Congressional Research Service, that became the Wilderness Act in 1964. The law established wilderness protection as a national policy by creating a national wilderness preservation system of lands under the jurisdiction of the Bureau of Land Management, the Forest Service, the U.S. Fish and Wildlife Service, and the National Park Service. (As of 2005, the Forest Service managed fifty-five wilderness areas that constitute forty percent of the National Wilderness Preservation System.) After passing the act, Congress immediately redesignated 9 million acres of national forest land as permanent wilderness areas from the 14 million the agency had already categorized as wilderness, wild, and primitive areas. That was out of the 186 million acres under Forest Service control. The act gave the Forest Service ten years to evaluate national forest areas that were without roads for wilderness status. The first evaluation, Roadless Area Review and Evaluation (RARE), received so much criticism when it was delivered in 1972 that a second report, RARE II, was ordered in 1977.

The Wilderness Act achieved the goals of the preservationists, but it contained substantial loopholes for development. The nine million acres immediately set aside had very little commercial timber, so the act did little to slow the timber harvest program. Although it prohibited roads, timber harvesting, and motorized vehicles, the act allowed water development, mining on valid claims, and commercial livestock grazing. Including grazing meant permitting fences, stock tanks, and maintenance activities on land that also had camping, hunting, and fishing. In short, conflicts over uses would continue—as would zoning land by federal legislation. Other acts that zoned the land include the National Trails System Act of 1968 and the Wild and Scenic Rivers Act of 1969, as well as the Endangered Species Act of 1966 with its 1969 and 1973 amendments.

Silent Spring and the 1960s

In 1962, in the middle of the final debates over the Wilderness Act, former U.S. Fish and Wildlife Service biologist Rachel Carson published *Silent Spring*. Carson presented a moving discussion of the impact of toxic chemical pesticides, primarily DDT, and their effects on the food chain and threat to human populations. She was highly critical of governments, agencies (including the Forest Service), and organizations for their failure to test chemicals in biotic settings, but was understanding of foresters and other resource managers who had learned the advantages of natural or biological controls of insects and other pests. Citing the work of ecologist

---

Wilderness Preservation

One of the issues the Forest Service has faced in managing wilderness is a contradiction between the provisions of the Wilderness Act and the very concept of wilderness. The designation of wilderness, environmental philosopher Mark Woods has noted, can legally sanction its destruction. Activities such as grazing, mining, and logging are allowed in wilderness areas if those activities predate wilderness designation. The law also requires some management activities. Furthermore, even passive use of the wilderness, such as backcountry camping, can degrade the land. Wilderness areas face many of the same threats facing other public lands: overuse, fire, invasive species, and pollution. Nonetheless, the Forest Service has to balance managing wilderness for the preservation of its naturalness with the opportunities for solitude and primitive recreation.

Source: Mark Woods, "Federal Wilderness Preservation in the United States: The Preservation of Wilderness?" in The Great New Wilderness Debate, *edited by J. Baird Callicott and Michael P. Nelson (Athens: University of Georgia Press, 1998), 131–53.*

Charles Elton, Carson "argued that diversity was the key to biological health. It was imperiled by the human conceit that sorted out wild species according to their human uses and eliminated the 'bad' ones."[346]

The book triggered a national controversy. Pesticide manufacturers and large agricultural organizations threatened lawsuits and attacked her credibility. Overlooked in the furor was Carson's call for research to determine how to use pesticides safely and to find alternate techniques for pest control; she had not urged the abandonment of pest control. Instead of following her suggestions, many in the timber and agricultural industries and the Forest Service spent the next twenty years and countless resources arguing that they could not carry out their work without the chemicals they had.

By the time Carson's book came out, the Forest Service was conducting aerial spraying of DDT on more than one million acres of national forest lands annually. The Forest Service was largely dependent upon pesticides and herbicides for generating high timber yields. By the late 1960s, moreover, scientific data on the impact of multiple uses of forest resources, most especially but not limited to logging, revealed that human activity was adversely affecting wildlife and habitat. Biologists and wildlife ecologists questioned the size of clearcut areas and predator programs on the national forests because of their impacts on animal populations. Meanwhile, recreational activities and overgrazing had affected land quality. The land could no longer sustain all activities all the time.

Forest Service leaders responded to *Silent Spring* with an "information and education" program in 1965 that called for publishing a book that would have the emotional impact of a *Silent Spring* while making the case for clearcutting. In another attempt to counter criticism, the agency asked several deans of forestry schools to examine its clearcutting program. Their findings largely exonerated the agency. Technical blunders had certainly occurred, it stated, but the deans saw no reason to discontinue the practice of clearcutting. The study, however, had little impact.[347]

Its effort to blunt criticism revealed an agency out of touch with mainstream thinking. Many foresters within the agency even advocated managing the land more intensively to achieve "full utilization." By 1970, the Forest Service was clearly on the defensive. Some in the agency believed they understood the source of the friction: the public was ignorant about clearcutting. Charles Connaughton, who served as the regional forester for three regions from 1951 to 1971, recognized this gap between the forestry profession and public perception. In an article published in 1966, he noted, "The toughest problem facing the forestry profession today results from a major segment of the public not realizing commercial forest lands can be managed without destroying its utility and appearance. Consequently, much of the public lacks confidence in foresters as stewards of the land." He encouraged his fellow professionals to adopt management objectives and techniques "which result in acceptable conditions on the land that the public can and should be shown." Four years later, he again wrote in defense of foresters' use of clearcutting

and faulted foresters for not doing enough to educate the public of its advantages. He exhorted foresters to show the public how clearcutting worked and was not harming the land, and to reconcile silviculture with aesthetics to win public support.[348] Connaughton's suggestions were too little, too late.

Public opinion had been largely indifferent regarding federal land management, but as historian Thomas Dunlap noted, "*Silent Spring* marked a watershed, as the private, scientific debate became a public, political issue."[349] The book gave impetus to Congress to pass the Clean Air Act (1963) and the Water Quality Control Act (1965), both of which have been amended and updated, and along with the Wilderness Act, helped usher in the modern environmental movement. Concerns over quality-of-life issues fueled the growing environmental movement, which underwent a change in perspective during this period. The human species no longer stood apart from the rest of the natural world, but was increasingly viewed as a part of it. Paradoxically, human survival was of growing concern. The constant threat of nuclear war, the escalating use of chemicals in Vietnam, and the confirmation of Carson's findings, combined with photographs of planet Earth taken from space, were reminders that despite technological progress, life was increasingly fragile. Eventually, all twelve of the most toxic agents Carson described in *Silent Spring* were banned or restricted.

The Forest Service used chemicals such as DDT and 2,4-D to control weeds in all sorts of settings and for all sorts of reasons. A ranger on the Davy Crockett National Forest was photographed using an herbicide to eradicate water lilies because, as the accompanying Forest Service caption read, "these plants may become a serious nuisance for boating and fishing." (USDA Forest Service)

As it had done in the 1930s to promote timber management, the Forest Service turned out pamphlets and booklets in the 1960s to educate the public about the benefits of clearcutting. Americans, however, were losing trust in the agency's motives and questioned the need for clearcutting. (USDA Forest Service)

The civil rights and antiwar movements in the 1960s led many people to question the establishment. Disillusionment with government policies deepened as the Vietnam War continued into the early 1970s and the Watergate scandal unfolded in 1973, and federal land managers were swept up in the general distrust. Some doubted whether the Forest Service could manage the land for the future when they learned of its continued clearcutting and use of chemical pesticides. "Ironically," as one environmental historian has noted, "the same deference for scientists that contributed to public acceptance of intensive management for maximum production in the 1950s now contributed to widespread questioning of the faith in technological fixes and a growing skepticism" toward the Forest Service.[350] An agency that had been used to thinking of itself as heroic now found itself characterized as villainous.

Foresters' scientific colleagues, especially wildlife biologists but also those working in other nontimber sciences, began to criticize Forest Service practices, too. A proposed timber sale on Montana's Lewis and Clark National Forest was opposed by wildlife biologists who found that logging activities, which were said to improve food and forage supplies, did not always yield sustenance for desired elk and deer populations. Furthermore, elk displaced from public land by the harvesting were forced onto private land where they had less protection. To deal with the issue, national forest managers began to work with federal and state wildlife agencies in the 1970s to conduct several long-term studies of elk habitat requirements. From these studies emerged a series of recommendations for designing and conducting timber sales to minimize their adverse effects on elk.[351]

In the early 1960s, forest managers on West Virginia's Monongahela National Forest faced concerns about creating large clearcuts in the management of hardwood forests. The practice was employed with increasing frequency in the eastern forests to meet demand for pulpwood. By 1969, even-age management accounted for about fifty percent of volume on eastern national forests.[352] But the large clearcuts angered hunters because of its effect on squirrel and turkey habitats in a highly valued hunting area on the Monongahela. Chief Edward Cliff dismissed their complaints as "a very self-centered protest from a very small segment of the population who wanted the national forest to be managed just for their own personal pleasure."[353]

In 1964, the West Virginia legislature commissioned a study of timber management practices and eventually asked for changes. For the next seven years, the Forest Service continued to practice even-age management on the Monongahela, with some modifications. The agency's own environmental impact statement supported the state commission's findings, but differences between the two sides remained.[354] In Montana, local citizens led by a former forest supervisor challenged the clearcutting and terracing on the steep mountains of the Bitterroot National Forest. Both issues became national controversies in the 1970s.

By the mid-1960s, with the Wilderness Act in hand, mainstream preservation organizations had the political clout and momentum to push Congress to pass additional legislation protecting waterways and endangered species. Several diverse groups—including scientists, philosophers, governmental officials, and conservationists—challenged the Forest Service's forestry practices and assumptions and maneuvered legislative constraints on its land management through Congress. But the commitment to producing timber remained the government's priority.[355] In the coming years, new legislation would democratize land management as never before.

The Spate of Environmental Laws

The Sierra Club and the Wilderness Society drew new members as environmental issues came to the fore in the 1960s. The Sierra Club expanded outside its traditional geographic region and became a truly national organization. Recreation groups experienced similar growth. No longer mere clubs, they were highly organized and professional outfits, and they soon added legal expertise to their scientific expertise. Their glossy magazines and books documented environmental issues and reported on their lobbying efforts on Capitol Hill.

In response to the growing environmental movement, Congress passed the National Environmental Policy Act (NEPA) in December 1969 and President Richard Nixon signed it into law on January 1, 1970. As one historian has summarized it, NEPA came from the recognition of "the profound impact on the environment of human activities, including urbanization, population growth, industrial pollution, resource exploitation, and technology. It asserted that humanity must promote the general welfare and bring 'man and nature' into productive harmony with each other, in order to provide for the needs of future generations."[356]

NEPA established the Council on Environmental Quality to monitor the performance of federal agencies in implementing NEPA and included creating a regulatory system through the preparation of environmental impact statements. The law stipulated that the federal government take "a systematic interdisciplinary approach to the integrated use of natural resources through the natural and social sciences."[357] It was intended to provide decision makers and the public with a rationale for choosing projects and to discuss the potential for environmental impacts and effects. Logging altered wildlife habitat and displaced animal populations, for example, and runoff from disturbed soil could affect watersheds and disrupt fisheries; the adverse results of logging must therefore be taken into consideration in management plans. NEPA aimed to merge the aesthetic and ecological considerations into an administrative whole.[358]

The law had a substantial impact on how the Forest Service assembled plans for managing the land. In addition to economic aspects of land management, the agency had to consider the aesthetic and scenic aspects of nature and report its findings. The law also required public disclosure and permitted citizen involvement in federal land management actions. The environmental impact statement quickly became a weapon in the legal arsenal used by environmental organizations to force the Forest Service and other agencies to examine from all sides the long-term implications of management decisions. Three hundred sixty-three NEPA-based lawsuits were initiated against federal agencies by March 1, 1973.

The Endangered Species Act (ESA) of 1973 arguably had an even greater impact on the way the Forest Service managed public land. Though the majority of lawsuits brought against the Forest Service have been for possible NEPA violations, the Endangered Species Act was the cause of the most dramatic changes in Forest Service planning. Research concerning the Forest Service's success in ESA cases between 1970 and 2002 suggests that the agency worked effectively to comply with the statute's mandates.[359]

The 1973 version of ESA greatly increased the authority and scope granted to a number of federal officers under previous endangered species legislation. Congress had already passed the Endangered Species Act in 1966, with amendments in 1969. The original act directed the U.S. Fish and Wildlife Service to prepare and maintain an official list of endangered native animals but provided no authority to regulate trade in these species and "takings"—the harming, harassing, removal, or destruction of a listed species on public or private land. In 1969, Congress gave Fish and Wildlife authority to expand the list to include mollusks and exotic species and gave the agency power to protect listed native species. The secretaries of Agriculture, Defense, and Interior were instructed to conserve and protect endangered species as well. The 1973 amendments further enhanced the government's ability to protect habitat for wildlife species threatened with extinction.[360]

The name of the act differs from the purpose. It is not to preserve endangered species, but rather to provide "a means whereby the ecosystems upon which endangered species and threatened species depend may be conserved, [and] to provide a program for the conservation of such endangered species and threatened species."[361] ESA protects habitats of endangered species and not the animals per se, a subtle difference with great implications for the Forest Service. Wildlife could not be removed to another location or raised in captivity, for example, and then brought back later when a new crop of trees had grown, as some legislators wanted to do.[362]

Under ESA, the secretary of Commerce has responsibility for most marine species, and the secretary of Interior, all other species, the definition of which was expanded to include all vertebrates or invertebrates, not just members of selected classes as in the 1969 act. The list was also expanded to include "threatened" species and thus protect species before they were in imminent danger of extinction. In addition, the agencies had to list animal populations within the species. The taking of species was explicitly prohibited in Section 9, further restricting where activities such as logging could be carried out.

Public participation in the listing or delisting of animals was encouraged. The law also allowed any person to bring action in U.S. district court for alleged violation of the act. The court then could prohibit any person or agency from acting in ways deemed harmful to endangered species. It was this part of the law that opened Pandora's box for the Forest Service. Max Peterson, who dealt with the problems created by the new law while chief from 1979 to 1987, noted that ESA left the Forest Service in a difficult position:

> We didn't have any real idea of the reach of the Endangered Species Act and I think there have been some real problems associated with the Endangered Species Act over the years because of like-equity considerations. If you've done a good job of managing your land and you've protected the habitat of species and your neighbor has not, and Endangered Species Act comes along, it's your habitat that gets designated as critical habitat. Because a lot of critters had their habitat changed, pretty soon the National Forest became the refuge area for a lot of endangered species. I don't think anybody understood that when the act passed. So the impact on the national forests was much greater.[363]

Dale Robertson, chief from 1987 to 1993, called ESA "the hammer" environmental groups used against the agency in the courts.[364] Every management plan had to take into account its impact on endangered or threatened species, and the agency had to coordinate its efforts with—and at times defer to—the U.S. Fish and Wildlife Service. If Fish and Wildlife decided that an animal's habitat was at risk because of Forest Service activity, that activity could be halted.[365] In the hands of lawyers from wildlife organizations trying to preserve critical habitat, ESA quickly became an effective means of stopping logging activities. In the 1980s, lawsuits filed on behalf of the red-cockaded woodpecker in the South and the northern spotted owl in the Pacific

Northwest forced the Forest Service to substantially alter its management plans and reduce timber harvesting to protect old-growth areas that were the birds' habitat.

To deal with the problems of budgeting year-to-year, Congress passed the Forest and Rangeland Renewable Resources Planning Act of 1974. Often referred to as RPA, the act tried to align congressional appropriations and Forest Service management objectives by requiring the agency to undertake long-range planning to give Congress a better idea of its funding needs. The Forest Service would now prepare ten-year assessments describing the nation's forest and rangeland renewable resources, prepare a five-year planning document assessing objectives over a forty-five year time frame, and submit an annual report.

RPA was intended to apply interdisciplinary planning to clear—by 2000—the backlog of programs and projects that had not received proper funding. Instead, the law generated more paperwork, which required the expansion of the agency's workforce, and it created the illusion of a shield from political criticism: if Congress or the White House refused to fund the entire program, the agency could point the blame elsewhere for its failure to meet its multiple-use obligations other than timber. Over the years, the White House and congressional representatives from timber states obligingly played the villain by threatening to slash the agency's budget if the Forest Service did not produce the amount of timber desired.[366]

Bitter Harvest on the Bitterroot

As early as the end of the 1960s, the Forest Service seemed under attack from all sides, and one former employee characterized the situation as "foresters on trial."[367] A few employees voiced criticism over clearcutting and other management decisions, knowing full well that they could be reassigned or fired for speaking out.[368] Meanwhile, twenty million people took to the streets for the first Earth Day celebration on April 22, 1970. The environmental movement was in full swing.

On the Bitterroot National Forest in Montana, the local population had become upset about the timber program's road construction and terracing of the hillsides, which residents likened to rice paddies in China. Orville Daniels, who became the forest supervisor on the Bitterroot in 1970, explained why people objected: "When you come around the corner and you look at it, you go, 'Oh, my God,' because it's high on a hill and the road comes right into the middle of it and the vista is right in the center of it and it is the most awful looking thing."[369]

Local hunters were concerned about the impact on elk populations. Shortly after Daniels took charge, a local banker asked that the Forest Service stop terracing and bulldozing the hillsides and cease clearcutting the forests. Daniels recalled, "He was as conservative and Republican a banker as you'll ever find, and he said, 'My wife is about to kill me if I don't get you to stop building the road across that mountainside over there. The Garden Club is upset.' The loggers were upset. It was their elk hunting, too."[370]

Conservation groups were not alone in criticizing clearcut areas. Former Forest Service forester Bob Wolf, who since 1953 had been working on Capitol Hill and either drafted or contributed to the writing or revision of nearly every significant piece of federal public land legislation, heard similar criticism from a citizen with a unique perspective on its impact in the 1960s:

> I was talking to a United Airlines pilot in Denver just by accident. And when I told him I was a forester, he really blasted me. He said, you know what, I used to be able to fly to the west coast and see nothing but solid forest. Now, he said, it looks like an area with smallpox, you know, all these clear cuts scattered all over. He thought that we foresters just didn't have it right. And that seemed of interest to me because this was a public opinion of a fellow who didn't come out of some conservation organization raising questions, you know; he just didn't like the way it looked.[370]

The bulldozing and terracing for timber production on the Bitterroot National Forest became known as the "Oh my God" clearcut. Although Chief Ed Cliff believed it to be "a forestry showcase," local citizens opposed it for several reasons, including the loss of hunting grounds. (USDA Forest Service)

Ed Cliff, the Forest Service chief from 1962 to 1972, when terracing and clearcutting on the Bitterroot became a national issue, later conceded, "It did look like a harsh treatment, and it created a drastic temporary change in the appearance of the environment." But, he continued, "I think the criticism of the practice from the standpoint of watershed protection was grossly exaggerated." Furthermore, he believed it was "a good conservation measure. I had an opportunity to examine some of the terraced areas on the Bitterroot in late 1980. And the regeneration looks

These clearcuts, made around 1957 on the Lewis and Clark National Forest in Montana, were twenty to fifty acres and designed to minimize the effect on watersheds. Such patches prompted an airline pilot to remark to forester Bob Wolf that the once-solid forests looked like they had smallpox. (USDA Forest Service)

wonderful. The effects of the terracing are largely obliterated by the growth of the young trees." He predicted that when the trees matured, the Bitterroot would be "a forestry showplace." Although he understood the environmentalists' complaint that clearcutting had "a very strong impact," he did not think that reason enough to do things differently.[372]

Senator Lee Metcalf of Montana, deluged with complaints about clearcutting, asked Arnold Bolle, the dean of the state's forestry school, to head a panel of experts to look at the agency's clearcutting and terracing activities in the Bitterroot National Forest. The Bolle Report brought the differences regarding timber management into sharp focus.

Timber and technology people believed the Bitterroot was the pinnacle of what intensive management could achieve on otherwise marginally productive land. Bulldozers cleared commercially worthless lodgepole pine and terraced the land so that a special tree-planting machine could put seedlings into the ground. Several decades later, the terraces would allow easy access for logging machinery to efficiently harvest the trees. Commercially "unproductive" natural forests were thus converted into "productive" forests producing a "timber crop." Such operations allowed agency timber managers to raise future lumber supply estimates, which, in turn, permitted them to increase the current allowable cut. It appeared to be a triumph of science over nature.[373] Opponents viewed the Bitterroot operations instead as a triumph of human arrogance and the placing of profits ahead of the long-term needs of the environment and the aesthetic sensibilities of humans.

Issued in December 1970, the Bolle Report provided a scathing indictment of the Forest Service's timber management on the Bitterroot. To compound matters for the agency, the report was issued around the same time as two internal Forest Service studies and a study by West Virginia. All four were highly critical. According to Forest Service historian Gerald W. Williams, Bitterroot employees ironically were angry over the Bolle Report but not the critical internal report, presumably because the former was made public.[374] The Bolle Report summarized the agency's overall behavior and policies over the past thirty years. After reporting that "multiple use management, in fact, does not exist as the governing principle on the Bitterroot National Forest," the authors wrote,

> *The heavy timber orientation is built in by legislative action and control, by executive direction and by budgetary restriction. It is further reinforced by the agency's own hiring and promotion policies and it is rationalized in the doctrines of its professional expertise....The rigid system developed during the expanded effort to meet the national housing post-war boom. It continues to exist in the face of a considerable change in our value system—a rising public concern with environmental quality. While the national demand for timber has abated considerably, the major emphasis on timber production continues.*[375]

The Forest Service was vulnerable. Multiple use was not its policy, and its defense of clearcutting seemed misguided when the study found that several large harvests were "entirely inappropriate, ruinous to future forest growth, and thoroughly destructive of values that were more important than timber."[376] These were essentially the same accusations the Forest Service had leveled at private industry for much of the first half of the twentieth century when trying to justify the agency's rationale for regulating lumbering on private land. Now the agency had to defend that same harvesting technique to justify not being regulated itself.

Legal action soon followed. Lawsuits were filed to stop clearcutting in West Virginia, Montana, and even Alaska. The lawsuits variously cited violations of the Multiple Use–Sustained Yield Act, the Wilderness Act, and NEPA. In the wake of the filings, the Senate Interior Committee's Subcommittee on Public Lands, chaired by Senator Frank Church, held contentious hearings in spring 1971 on timber and clearcutting. The Church Committee, as it became known, issued twelve well-received guidelines in three areas: timber harvest levels, use of clearcutting, and the environmental content of timber sale contracts. Congress later incorporated the Church guidelines in the National Forest Management Act of 1976, but in the meantime, both the Forest Service and the Bureau of Land Management agreed in April 1972 to abide by the guidelines for the federal lands under their control. Some thought the agreements marked a truce in the clearcutting controversy. If it was a truce, it was very brief.[377]

Gifford Bryce Pinchot, the only son of the first chief and a founding member of the Natural Resources Defense Council, visited the Bitterroot when his organization was looking for a suitable forest on which to initiate a lawsuit against clearcutting. When he saw the clearcuts there in summer 1972, he exclaimed, "If my father had seen this, he would have cried."[378]

OCALA NATIONAL FOREST
MANAGED FOR MULTIPLE USE OF FOREST RESOURCE

WATER FOR HOMES, INDUSTRY, AND AGRICULTURE

TIMBER FOR JOBS AND OUR NATION'S NEEDS

WILDLIFE FOOD AND COVER FOR GAME AND

RECREATION OUTDOOR ACTIVITIES FOR EVERYONE TO

FOREST SERVICE
DEPARTMENT OF AGRICULTURE

The Forest Service conducted tours and open houses to counteract unfavorable public reaction to multiple use and clearcutting on national forests. Max Wallace explained multiple use to Ocala National Forest visitors in 1962 (above), and Deputy Forest Supervisor Merwyn Reed described even-age management to a group of forestry school deans on the Monongahela National Forest in 1968 (facing page). (USDA Forest Service)

Indeed, his father may very well have cried or, more likely, "roared with wrath," since he had spent his career fighting against clearcutting.[379] To the son, this was evidence of how far the Forest Service had strayed from the vision of the first chief, who had favored selection cutting to conciliate "the doctors, sportsmen, and practical men."[380] Al Wiener, an appraisal specialist with the Forest Service's Division of Timber Management, however, rebuked the son's contention by declaring that "Gifford Pinchot would have laughed" at such statements, and implied that Bryce Pinchot—and by inference, those who thought like him—was a "dilettante amateur" who had no business judging the work of foresters.[381] Such was the anger within the agency unleashed by the Bolle Report.

The Forest Service and lumber interests, meanwhile, found support in the White House. President Nixon appointed the President's Advisory Panel on Timber and the Environment in September 1971, which issued its report in April 1973, after the Bolle Report came out. The panel defended the status quo, and its twenty recommendations in one way or another urged further development of the land to improve the productivity of the national forests. It offered qualifying statements about preserving environmental quality and protecting nontimber uses of the land to give the appearance of balance and concern, but it did so almost as an afterthought. At a time when more than half the 186 million acres of the national forests was already classified

as "commercial timberland," the president's panel concluded that cutting in old-growth forests could be accelerated from fifty to one hundred percent above current levels.[382]

External pressures and internal rigidity had combined to delude Forest Service leaders into thinking that meeting timber quotas that satisfied Congress and private industry was sufficient. The Bolle Report was a wakeup call. Chief Ed Cliff expressed concern about the direction in which the Forest Service was heading and believed that the agency's programs were out of balance,[383] but he faulted the budgetary process: Congress and several presidential administrations had enthusiastically supported programs that generated revenue over those that did not. Though the Forest Service never received as much funding for non-timber programs as agency leaders said they wanted, they never pressed too hard for fear of that Congress would slash its budget.[384]

While the Bolle panel was examining the Bitterroot work, the Izaak Walton League had joined with two turkey hunters—coal miners whom Chief Cliff characterized as "self-centered" and "militants"—to file suit in 1973 to stop the clearcutting on the Monongahela National Forest. The suit alleged that in undertaking clearcutting, the agency had violated both the letter and the intent of the Organic Act of 1897, which required the marking of dead and mature trees for harvest; by cutting all trees—young as well as mature—without marking, the Forest Service had exceeded its authority. The judge agreed and urged the agency to go to Congress to have the law changed to meet standards of modern forestry. The decision, which the appeals court upheld in 1975, temporarily shut down timber operations in the region. Similar suits were pending, and the Forest Service faced the real possibility of having timber operations shut down around the country.[386]

With that ruling, the overemphasis on timber during the 1950s and 1960s finally came back to haunt the Forest Service. The assumption that the agency had the right to reinterpret the law as times changed failed to take into account that the needs of the public might also change. Cliff's characterization of the West Virginia turkey hunters as militant in their beliefs could equally well have described the agency in the 1970s. The Forest Service's refusal to listen to opposing viewpoints regarding clearcutting had angered environmental advocates and ended the good relations between the Forest Service and the general public. As Bob Wolf described it, "The love affair started ending in the '60s, but the divorce didn't occur until the '70s with the Monongahela and the Bitterroot," yet "the Forest Service kept thinking that they could patch things up and they couldn't."[387] Following those two legal decisions, the Forest Service began a twenty-five-year odyssey in the leadership wilderness in which it struggled first to retain and then to regain control of its land management mission.

National Forest Management Act

The government decided to heed the judge's recommendation and went to Congress for legislative relief before other pending lawsuits shut down all Forest Service timber operations. Senator Humphrey sponsored compromise legislation. After several months of negotiation between those who wanted the Forest Service "to retain the widest possible latitude in forestry practices and those wanting to rein in the agency through highly prescriptive legislation," Congress produced the National Forest Management Act (NFMA) in 1976.[388]

This act amended the Resources Planning Act, repealed much of the 1897 Organic Act, and gave the Forest Service a new organic act. It mandated a nationwide, forest-by-forest planning process with emphasis on "land-management planning, timber-management actions, and public participation in decision making. It set forth broad policies for the national forests, including some specifics that had formerly been discretionary," but it allowed the Forest Service to retain its professional judgment in forest management, including harvesting. Clearcutting could continue, with some limitations. The act did not change what the Forest Service did on the ground, but how it planned to do it. This "refined endorsement of multiple use" moved the arguments out of the woods and into the planning process, ensuring that the agency, in the words of David Clary in his study of the Forest Service and timber, "would never be relieved of controversy."[389]

As a supplement to RPA, NFMA specified the need to develop integrated land and resource management plans for each national forest and grassland every ten to fifteen years, identify areas best suited for timber production, and monitor the effects of management. From NEPA came the requirements of public participation and interdisciplinary analysis. A committee of scientists helped devise regulations to implement the law. They were initially issued in 1979 and revised again in 1982. One of the regulations was that the Forest Service would maintain well-distributed, viable populations of all native vertebrates in national forests and grasslands. This requirement led to the hiring of scientists from nonforestry fields, such as biology, hydrology, and soil science, and it fundamentally altered how the Forest Service viewed and dealt with forest resources.[390]

Another reform was the adoption of a restrictive definition of sustained yield called nondeclining even-flow. Section 11 of the act limited the sale of timber from each national forest to a quantity that could not exceed the amount of timber removed annually in perpetuity on a sustained-yield basis. This maintained the illusion of sustained yield, but exceptions—for salvage logging in the case of fire, disease, windthrow, or other natural disturbance—allowed the Forest Service and timber industry advocates to exceed the apparent limits.

Instead of dealing with rigid annual harvest limits, each forest would adhere to allowable sale quantities of timber (ASQ), the average annual amount of timber that could be sustainably harvested. ASQ is a maximum—not a target or goal—and cannot be exceeded over the ten-year forest planning period. The Forest Service could increase ASQs if "intensified management practices, such as reforestation, thinning, and tree improvement" would produce greater harvests in the future. Thus the agency could harvest more timber immediately in anticipation of future growth that it could not guarantee, but if at the end of the ten-year cycle, timber volume did not meet expectations, forest managers were supposed to reduce harvest levels.[391]

No sooner had Congress finished passing the landmark environmental laws of the 1960s and 1970s than the Kaibab Game Preserve disaster resurfaced to offer a timely example of why integrated management was important to forest health. In the 1970s, ecologists examined what had happened in the 1920s on the Kaibab and began questioning the assumption that predator control had permitted the deer herd to expand beyond the carrying capacity of the land. Information that would support such a conclusion was missing, faulty, or incomplete, and some data had been overlooked: there was no proof that predators had originally limited the herd, and the predator policy established afterward had no scientific basis. What had been accepted was rejected, just like the necessity of even-age management. Henceforth, the Forest Service would have to give more careful consideration to both science and changing public values.

AMERICAN

FORESTS
The Magazine of Forests, Soil, Water, Wildlife, and Outdoor Recreation

To Regenerate Eastern Hardwoods - - - -
CLEARCUT

by William E. McQuilkin

REPRINTED FROM AMERICAN FORESTS

June 1970

The Magazine of
The American Forestry Association
919 17th Street, N. W.
Washington, D. C. 20006

By the end of the 1960s, organizations that had traditionally been supporters of the agency, like American Forests, were openly questioning the practice of clearcutting. This article, written by a Forest Service researcher in defense of clearcutting in West Virginia, appeared in a 1970 issue of *American Forests* alongside several articles criticizing clearcutting. (USDA Forest Service)

Chapter Seven

NEW FACES, CHANGING VALUES

Smokejumper Jeanine Faulkner (right), Lolo National Forest, 2000. (USDA Forest Service)

As the Forest Service grew in size and responsibility during the 1950s and 1960s, it achieved the height of its popularity, even as its policies began to come under fire and legislation forced changes in its operations. Some of the agency's difficulties in the decades that followed were due to the makeup of its personnel. New environmental laws and procedures required adding nonforestry scientists and experts to the planning teams. Women, long relegated to office work or unpaid support positions since the agency's founding in 1905, fought for and won the right to work in field positions. African Americans, Native Americans, and Hispanics soon joined them in moving into those positions and up the ladder into administrative jobs.

It's a Family Affair

In the early years of the Forest Service, rangers' wives provided a convenient and much-needed source of free labor on understaffed and underfunded ranger districts. Wives sorted and date-stamped the mail, prepared correspondence and reports, and operated switchboards. They also made and delivered meals to fire fighters and sometimes stayed to help fight. In one extreme example of familial dedication to the agency, Harriet Eveleth, a ranger's wife, purportedly left her three small children behind during the Big Blowup of 1910 and rode twenty miles on horseback from Monarch to Neihart, Montana, to take charge of a hundred men fighting a fire, not knowing whether the fire would jump the ravine and claim her home and family. Eveleth's bravery earned her the nickname "The Paul Revere of the Belts."[392]

Other wives took on less dramatic roles in time of crisis. They served as nurses and camp cooks on the fire lines, helped prepare fire reports by estimating timber and forage losses, and finally analyzed the causes of the fires. When Washington refused to send more equipment because records showed forests had their allotment, wives dug into their personal finances to pay for tools. Families also hosted visiting stockmen, hunters, or forest supervisors overnight in their homes. Supervisors and administrators viewed women who declined to take on the role of active helpmate as disloyal and hindrances to their husbands' careers, and rangers with uncooperative wives tended not to stay in the Forest Service very long.[393]

Several wives saved their husbands' careers because of their behind-the-scenes contributions, something the Forest Service quietly acknowledged at times. Emma McCloud, wife of forest officer Mal McCloud, handled most of her husband's correspondence during his thirty years of service and curbed his "weakness for liquor" to keep him employed. Charles Shinn was on the

Often shorthanded while fighting fires in the early years, rangers' wives like these two on the Sierra National Forest provided critical assistance, even leading fire crews when necessary. (USDA Forest Service)

brink of losing his job on the Sierra when his wife Julia became his paid clerk. Her office managerial skills complemented his field abilities, and despite rules against nepotism, the need for good field leaders was such that Shinn's superiors kept him on as long as his wife worked for him. When Charles retired in 1911 after a long and distinguished career, Julia continued serving as a clerk in the Sierra office until 1923. Rangers relied on her professional advice so heavily that her husband's replacement had to instruct his men to come to him—not Julia—to discuss their problems: he could not establish himself as supervisor as long as they looked to his clerk for leadership and advice.[394]

Though most wives were not compensated for their work, they faced scrutiny and criticism from their husbands' superiors. As in the military, wives were expected to toe the line and be supportive if their husbands wanted to succeed in the organization. Retired forester Bob Wolf recalled the expected role of wives during the 1950s and 1960s:

When I got my first efficiency rating, it was excellent except to say that my wife wasn't very sociable. The ranger's wife had teas and my wife didn't go to them. She was working in the hospital, which was sixty miles away. And when she saw that, she went down to the ranger's office and demanded that she be paid, or else take her name out of the efficiency report. It was a very paternalistic operation, you know. So a ranger's wife was a sort of a queen bee and all the women in the ranger station were supposed to march to her orders.[395]

Like the military, the agency moved personnel around every few years, which required families to adjust to a new place and the husband's new boss. In contrast to the Wolfs' experience, Marian Leisz, wife of former forester Doug Leisz, found that fellow Forest Service wives eased the adjustment of moving and fostered a feeling that the Forest Service was a large, extended family:

The wives in general were hostesses on the ranger stations. And so we would have potluck picnics probably once a month in the summertime. And sometimes this was done on a forest level so that all the ranger districts on the forest would get together for a picnic. They would include all the young guys that were out there working in the summertime, you know, that were just bunking in bunkhouses and include them and that was lots of fun, too. And when people came through from the regional office or from an adjoining ranger district, you know, they'd always stay for dinner or one of those kinds of things. It was just expected….

All of us had big families and on a ranger station there might be, you know, three families and fifteen kids. Families were very important in our lives….And living on a compound like that had its wonderful advantages, and most of the time just wonderful relationships came out of it. After about

the first or second move, every time we moved there was someone there we had known earlier. And, so we did have an extended Forest Service family all through our career.[396]

"Family" in Forest Service vernacular even carried over into the workplace. As early as the 1910s, Forest Service employee meetings have been called "family meetings." Ellie Towns, an African American woman who started with the agency in 1978 and retired as southwestern regional forester in 2002, was initially skeptical:

The look of the Forest Service family changed dramatically between 1963, when the Stanislaus National Forest rangers posed for a photograph (top), and 1978, when an employee civil rights meeting was held. (USDA Forest Service)

That word "family," it bothered me when I first came into the Service because it seemed to me that everybody knew everybody else but I didn't know any of them. And we'd go to meetings and there were all these relationships and these friendships and I thought, "Gee, I'll never be a part of that." And I was thinking about that near the end of my career, that word "family," and of course now everywhere I go in the Service, even now that I'm retired, I know people and I do have that sense of family. You know, we have family meetings; we don't have employee meetings. And I used to think, ooh, why do they call them that, why don't they just call them employee meetings, which is what they are? Well, now I feel really warm and fuzzy about that.[397]

The Can-Do Agency and the Mythical Ranger

In the 1950s, the Forest Service reached its peak in power and prestige and was the undisputed leader in American conservation. Chief Silcox's success in suppressing forest fires gave only an inkling of what could be accomplished if the agency applied the right mix of manpower, science, and spirit. As former Chief Michael Dombeck (1996–2001) described it, "If commercially valuable timber was inaccessible, build a road. If harvested forest on south-facing slopes resisted regeneration, terrace the mountainside. If soil fertility was lacking, fertilize the area. If pests or fire threatened forest stands, apply pesticides and marshal all hands to combat fire. If people grew unhappy with the sight of large clearcuts, leave 'beauty strips' of trees along roadways to block timber harvest units from view."[398]

The Forest Service soon became known as the "can-do" agency, the Marine Corps of the civil service. The comparison to the military was apt. Pinchot had patterned the agency's structure and organization in part on the somewhat paramilitary Prussian forest service, and the early uniform designs mimicked those of the U.S. Army. For many in the Forest Service, military service during World War I was a source of great pride. Men who had served in the forest engineers, including Henry Graves, William Greeley, and Robert Stuart, preferred to be addressed by their military rank afterward, even when they were chief. Evan Kelley, who also served as an officer in the forest engineers during World War I, went by "Major Evan Kelley" until he retired in 1944.[399]

World War II reinforced the military connection in the Forest Service. In addition to recruiting smokejumpers for paratrooper service, the Army recruited Forest Service rangers for the 10th Mountain Division. The division trained at Camp Hale, which was built on land the Army acquired by permit from the White River National Forest near Leadville, Colorado. On the steep hillsides adjacent to the camp, situated above 10,000 feet, soldiers learned skiing, rock climbing, and winter survival skills in temperatures that often dropped to thirty degrees below zero. After the war, several men from the division became involved in developing the ski industry in western states carved from national forests; others served as "snow rangers" in the Forest Service. In 1965, the military turned Camp Hale back over to the Forest Service.[400]

Time in the military taught the men that there were few obstacles they could not overcome. They learned to work within the limits of military regulations and yet remain flexible and adaptive to get things done. After the war, soldiers used the G.I. Bill to go to forestry school and then joined the Forest Service. Veterans were attracted to smokejumping and firefighting and brought a military bravado to those jobs, and conversely, the discipline and excitement of firefighting and smokejumping inspired some Forest Service men to join the military. The command-and-control administrative style of the military that had typified Forest Service cultural and administrative behavior since World War I became more ingrained and made Forest Service employees both conformist and inward looking.

The foresters and engineers who dominated leadership positions came from similar backgrounds. They were white males, usually from middle-class families and rural, conservative backgrounds. They trained in one of twenty-seven forestry programs that all emphasized timber production yet required little if any understanding of nontimber resources.[401] Those with military experience were unlikely to question authority and placed the interests of the agency above their own.

A 1949 study of Forest Service administration, conducted by the U.S. Army, praised the agency as "representative of many of the finer principles we associate with the American way of life." It extolled "the democratic way in which relationships are handled, the dedication to the worthwhile

After World War II, alpine skiing became enormously popular in the United States, in part because of Forest Service efforts. As of 2005, 135 alpine ski areas had been built on national forests, including Arapahoe Basin in Colorado, where this image was taken in June 1998. (Author's personal collection)

concept of conservation, the continual striving for efficiency and effectiveness of job accomplishment and an organizational morale second to none."[402] It is little wonder that the Army found much to like in the Forest Service: supervisors with military backgrounds often used fear and intimidation to motivate their men. Edgar Brannon, a landscape architect who entered the Forest Service in the early 1970s, remembered their style of leadership:

This post-World War II leadership culture, these were strong people in charge—they chew tobacco, they drink whiskey, they smoke cigarettes, they are hard. They are tremendously dedicated, but they are, in some sense, a bunch of characters, too. The leadership style is really command, control, [and] intimidation. Basically, if the people that work for you don't fear you, then you're "country clubbing" it—you know, you're not getting the most out of your people. That's the style. I can remember an early forest supervisor I worked for who basically began his day drinking coffee and reading the morning paper, and then at ten o'clock, he would begin walking down the hall, and I could hear the staff officers begin to tremble because he was, as he would say, looking for someone's ass to chew. And he would go in and pick somebody—seemed to be at random but I'm sure it was focused—and basically ream them out. There was a sincere dedication from these people to really do the job and they were so into the Forest Service, it was their entire life.[403]

In 1960, Herbert Kaufman published a study of administrative behavior in the Forest Service. He sought to learn how field personnel operating under the agency's decentralized system, which allowed the lowest-ranking officers to make decisions without consulting superior officers, succeeded at consistently high levels. Kaufman found that the Forest Service recruited men with technical knowledge and practical skills who also had the will to conform and carry out what he called "the preformed decisions" of their superiors, which could be found in the ranger's bible, the *Forest Service Manual*. No longer a slim volume that fit in a shirt pocket, the manual had become a multivolume set of loose-leaf binders. The agency designed the manual to do most of the thinking for rangers: decisions on everything from "free-use permits to huge sales of timber, from burning permits to fighting large fires, from requisitioning office supplies to maintaining discipline, classes of situations and patterns of response" were detailed in the manual. The manual and agency culture ensured a standard way of handling the situation or problem, regardless of where it occurred.[404]

Rangers also kept diaries and filed reports that would eventually reveal deviation. Because personnel were rotated every two to three years, any inconsistencies might be found and reported by one's successor. In such an atmosphere, a forester who questioned operations might be labeled a troublemaker and place his career at risk. By handling personnel this way, Kaufman noted, the Forest Service "enjoyed a substantial degree of success in producing field behavior consistent with headquarters directives and suggestions."[405]

According to Forest Service Historian Gerald W. Williams, the agency's mobility policy benefited its employees as well: it gave them broad experience in managing different resources and different people. Mobility was a prerequisite for advancement; it screened and trained future leaders and gave them a national perspective on the agency. It also weakened ties to one community, ranger district, or forest so that decisions would be based on national priorities. The effectiveness of mobility, however, fell victim to changing circumstances. Some chiefs wanted more movement, while others wanted less—sometimes for budgetary reasons. A tight budget might mean there was less money available for moving families, and so they were moved less often.

Within the agency, there may have been disagreement about what to do or how to do it, but once a decision was made, everyone accepted it and worked to implement it.[406] That a forester's peers rarely questioned his decision contributed to a sense of always doing what was best for the land. The emphasis on conformity and obedience fostered what one forester called the "myth of the omnipotent forester," an attitude that came to dominate the agency's thinking. In the mid-1960s, a seasoned forester told newly hired foresters, "We must have enough guts to stand up and tell the public how their land should be managed. As professional foresters, we know what's best for the land."[407]

Through its ever-increasing timber yields, the Forest Service was making tangible contributions to the growing U.S. economy and the struggle against communism—no small motivation for employees of a goal-oriented agency in the Cold War. In the age of Sputnik, scientific achievement mixed with a can-do attitude made the Forest Service a model agency and transformed the forest ranger and his partner, Smokey Bear, into iconic figures. Forest Service rangers wore the proverbial white hat, an image the agency had cultivated since the early 1930s through a radio program, "Uncle Sam's Forest Rangers," which had aired on Thursdays at lunchtime from 1932 to 1944 as part of the NBC radio network's *National Farm and Home Hour* program. The drama taught housewives and children about the agency's range, logging, and fire policies, and that well-mannered rangers on their rare visit to the city always removed their hats in the presence of ladies, a courtesy ignored by callous city dwellers.[408] After the war, countless agency-approved fiction and nonfiction books, and various movies and radio and television programs portrayed the forest ranger as the epitome of the mid-twentieth-century American man: "The man getting out of the station wagon was tall and well built. The forest ranger uniform with its big Stetson seemed a part of the man himself. His friendly, yet strong face appealed to [the young protagonist], who found himself liking the man at first sight."[409] Women wanted to be with him and men wanted to be like him.

The typical narrative found the forest ranger, who personified the agency, living an exemplary life in the woods while carrying out a job that brought him into conflict with people abusing the land. The confrontations presented an opportunity to educate the young audience. "Multiple use," at first merely implicit in the early radio program and in books in the 1950s, became a

The Forest Service had effectively used radio programs since the 1930s, but in the 1950s it began cooperating with Hollywood to promote its mission and the dedication of its employees through television shows like *Lassie* and films like *The Forest Rangers*. (USDA Forest Service – Forest History Society)

repeated mantra by the 1960s. When the collie Lassie, a character already popular from movies and television, joined Ranger Corey Stuart for a series of adventures in the 1960s in authorized books and a television series, the forest ranger reached the peak of his popularity.[410]

Yet despite the agency's efforts to promote itself, the public constantly confused the Forest Service with the National Park Service as well as the purpose of each agency. Even Smokey wore a Park Service–style hat. Such war campaign hats, with the pinched peak, had not been worn by Forest Service personnel in more than twenty years. The confusion only deepened as civic groups and state organizations began linking Smokey to broader conservation issues, such as fighting pollution, commonly associated with other federal agencies.[411]

Women: The Deskbound Years

Women had worked in clerical positions as "typewriters" in the Washington headquarters office since the agency's Division of Forestry days. Before World War II, the agency hired very few women for professional positions. Eloise Gerry, the first woman appointed to the professional staff of the Forest Products Laboratory, just after its opening in 1910, is a note-worthy figure not only because of her scientific achievements but also as an exception to the men's-club attitude that prevailed well into the late twentieth century. In the 1910s, the agency began hiring women as draftsmen, bibliographers, and what would later be called information specialists but made it clear that women were not welcome to apply for jobs that took them into the field. That remained the agency's position until the 1970s.[412]

During Gerry's career, the agency made one notable exception to its position regarding women in the field. With the fire season of 1913 approaching in northern California, Assistant Fire Ranger M. H. McCarthy wrote to his boss, Klamath Forest Supervisor W. B. Rider, to inform him that last year's fire lookout would not be returning to Eddy's Gulch Lookout Station because he had found a better-paying job. McCarthy had three applicants to submit for review. McCarthy thought so little of the first applicant that he

The Difference Between THE FOREST SERVICE AND THE PARK SERVICE

Forest Ranger

Park Ranger

The National Forests are lands of many uses. In them, the lands are managed to produce water for towns and cities, cattle are grazed, timber is cut for market, hunters and fishermen are welcome, skiing and camping are encouraged. Forest Rangers manage the National Forests. Lassie and I work for the Forest Service, which, as I have already said, is part of the U. S. Department of Agriculture.

One way you can tell a Park Ranger from a Forest Ranger is by his uniform. There is a difference, especially in the hat, badge, and shoulder patch.

The National Parks, such as Yellowstone and the Grand Canyon, are great outdoor museums. They are managed by Park Rangers. The only roads, trails, and buildings that are permitted in these public lands are those which they need to protect and manage them and provide for the comfort of visitors. National Parks specialize in preserving nature and do not produce crops as do the National Forests. The National Park Service is in the Department of Interior.

The Forest Service has long struggled to differentiate in the public's mind its multiple-use mission from that of the National Park Service's "single use" mission. In the 1960s, when the Park Service was pushing its "Mission 66" program, this page appeared in *The Forest Ranger Handbook with Corey Stuart and Lassie* interpretive booklet, produced for children with the agency's cooperation. The description touts national forests as the "lands of many uses" and points out the differences in uniform styles. (Personal collection of Sandra Forney)

declared bluntly, "I could not conscientiously recommend him, even in a 'pinch.'"[413] Though the second applicant had poor eyesight, it did not prevent him from frequently violating the local game laws.

"The third applicant is also 'no gentleman,'" McCarthy continued, but would nonetheless make a "first-class Lookout." McCarthy's suggestion was so unprecedented, he warned Rider, it "may perhaps take your breath away, and I hope your heart is strong enough to stand the shock." He recommended Hallie Morse Daggett, "a wide-awake woman of 30 years, who…is absolutely devoid of the timidity which is ordinarily associated with her sex as she is not afraid of anything that walks, creeps, or flies. She is a perfect lady in every respect, and her qualifications for the position are vouched for by all who know of her aspirations."

McCarthy urged his supervisor to try "the novel experiment of a woman Lookout." He also told Rider not to worry about being overrun by female applicants in the future "since we can hardly expect these positions to ever become very popular with the Fair Sex." McCarthy's faith in Daggett proved justified. She was one of the most effective lookouts on the Klamath National Forest, typically reporting fires before others did. Of the approximately forty fires she reported that first season, fewer than five acres burned. She also garnered national publicity, and the Forest Service soon had its choice of women for lookout positions. One forester optimistically predicted, "We may have [in] some time not only female forest guards but female forest rangers and even supervisors."[414] Indeed, by 1920, women had applied for jobs as rangers and grazing assistants, but the agency turned away their applications.[415]

Like all lookouts, Daggett had a telephone on which her supervisor called three times a day to check in. Every day she climbed a twenty-foot pole to take weather readings in winds up to fifty miles per hour. Daggett had a relatively easy time with other women lookouts. She had a log cabin, and once a week, her equally rugged sister made the six-hour round trip to deliver mail and provisions. Other lookouts might receive visits every two weeks and typically lived in more primitive conditions. In Washington, Colville National Forest lookout Gladys Murray reported that the "Forest Supervisor kindly approved my request for a warm log cabin for next year and the sturdy building is now ready to roof and receive its windows." She also had to haul drinking water from a spring twenty minutes' ride away and could do so only between dusk, when visibility dropped, and nightfall, when fires would again be visible.[416] Lookouts took in stride such difficulties as sunburn, scorpions, mice, rats, bears, coyotes, and high winds that blew down their tents.[417] Ironically, if a fire was nearby, agency rules required the female lookout to call a male smokechaser to come fight it, even if it was threatening

Hallie Morse Daggett (below), the first woman fire lookout, served for fifteen years. Helen Dowe had a more diverse career than most agency women of her generation. From 1919 to 1921, she served as a fire lookout on the Pike, in Colorado, before she married J. Burgess, who was in charge of maps and surveying for the Rocky Mountain region. Dowe (pictured with her husband) then worked on his survey crew before becoming a topographer with the agency. (USDA Forest Service)

the lookout post. Most women ignored the rule, since help could be several hours away.[418]

Serving as a clerk provided the other major opportunity for women in the Forest Service. Before Chief Pinchot reorganized the Forest Service and established regional offices in 1908, women rarely worked in the forest supervisor's office. The reorganization created new jobs and the opportunity to move west. Initially, men deemed the work too rough for women, contending it required a "two-fisted ranger" or forest officer to assemble and ship fire tools, round up volunteer firefighters from bars and saloons, and perform other nonclerical tasks. As the men advanced, however, women found themselves tackling the work of the "two-fisted ranger" as well as paperwork. Office work quickly became a "pink collar" job.[419]

A district clerk was the backbone of the organization, providing continuity between district rangers as they rotated through and briefing the new rangers on local issues. During her forty-six years with the Forest Service, Gertrude Becker worked for ten district rangers and ended her career on a note of triumph: "The last one, it took me twelve years to get him squared away. He'll tell you that. But we made it."[420]

Clerks took care of expected clerical duties such as payroll, issuing permits, and hiring seasonal employees, and worked as much as eleven hours a day five days a week. With the ranger often in the field, the clerk also became the public face of the Forest Service. Clerks "had to be schooled in what the agency was all about" to interact with users of the national forests—ranchers, miners, loggers, or vacationers—concerning rules, regulations, and local conditions. It became agency folklore that the district clerk of the 1950s and 1960s did the job of twelve people today.[421]

The Forest Service did hire thirteen women with forestry degrees before World War II, but they remained deskbound, prevented from doing the ranger's rough-and-tumble job in the field. In 1934, the Forest Service appointed Alice Goen Jones as an entry-level junior forester in Region 5. Jones had a degree in forestry from the University of California at Berkeley, but the agency's position on women as forest rangers had been made clear three years before her appointment in *The Forest Rangers' Catechism in Region Five* (1931): "Women are not appointed by the Forest Service as members of the field force even if they pass the civil service examination." Jones remained in research throughout her career and, as late as 1972, still encountered sexual discrimination.[422]

World War II temporarily allowed women to get out from behind their desks and demonstrate their field skills. In addition to Forest Service positions such as fire lookout and patrol, cooks for fire crews, telephone operators, patrolmen, and truck drivers, women took over traditionally male jobs in private industry—logging, operating mill saws, and scaling lumber. But when the war ended, women were removed from their jobs in favor of men returning home. The end of the war also spelled the end for the old-style ranger who had gotten the job because he lived in the area and knew the land and his neighbors. After World War II, as land management became more professional and complicated, a ranger needed to have a college degree. The G.I. Bill enabled veterans to go to college and earn degrees in forestry.

After World War II, the Forest Service continued to discourage women from applying for junior forester positions. Officials held to the old assumption that a female forester would get pregnant and resign to start a family or subordinate her career to that of her husband and move away. And if she married a forester, nepotism laws required one of them to leave the Forest Service.[423]

An agency employment leaflet from around 1950 stated the agency's position on women in field positions: "The field work of the Forest Service is strictly a man's job because of the physical requirements, the arduous nature of the work, and the work environment."[424] The only way to find out whether women could do the job was to hire them, but that was not permitted: it was a man's job. The Civil Rights Act of 1964, which required employers to provide equal employment opportunities, meant the agency would have to change its hiring practices.

That same year, President Lyndon Johnson signed legislation creating Job Corps. Part of Johnson's slate of "Great Society" programs, Job Corps gave job training and skills, along with some basic schooling, to unemployed young men (women were admitted later) from deprived backgrounds who were not in school. Although Job Corps workers participated in firefighting, community work, and forestry activities on the national forests, not all Job Corps work focused on conservation projects. In 1970, Congress established the Youth Conservation Corps expressly for that purpose. Like the Civilian Conservation Corps on which it was based, the YCC did not offer job training but it did give young men and women summer employment opportunities while giving the Forest Service the workers to deal with its backlog of conservation projects. These and similar programs exposed both rural and urban young people to conservation work, and some participants later joined the Forest Service because of their experience.[425]

The feminist and civil rights movements were slow to affect the Forest Service. As late as 1976, women held eighty-four percent of clerical jobs in the agency and fifteen percent of administrative and technical jobs, but fewer than two percent of full-time professional jobs.[426] The career of Geraldine "Geri" Bergen Larson was typical of the handful of women with a forestry degree. Although she ranked at the top of the 1962 forestry class at Berkeley and then earned a master's degree in botany, Larson had to work in research and public information

Job opportunities for women were limited largely to office and clerical tasks until passage of the Civil Rights Act of 1964, which required employers to provide equal employment opportunities. Several more years passed, however, before women began working in field positions. (USDA Forest Service)

instead of in the field, as she hoped to do, from 1967 to 1972. Her work on environmental issues and her educational background led to her appointment as the regional environmental coordinator for Region 5 in 1972, an unusual position for a woman to hold at that time. She developed regional policy to implement the National Environmental Policy Act, consulted in the field with people working on environmental impact statements, and coordinated those and other similar activities with the Washington office and other federal agencies.[427]

Larson still wanted to work in forest management. Bob Lancaster, the forest supervisor on the Tahoe National Forest, discussed her aspirations with Doug Leisz, the regional forester. Leisz hesitated because Larson's husband, who owned his own business in San Francisco, would have to move in order for her to advance in the agency. She and her husband worked out a compromise that allowed her to accept the appointment as deputy forest supervisor of the Tahoe National Forest in 1978, making her the first female line officer. She took over the Tahoe in 1985 and became the first female forest supervisor in the agency's history.[428]

A year after Larson made it into the field as deputy forest supervisor, the first woman candidate for smokejumper training arrived at McCall smokejumper base in Idaho. Women were not hired on a permanent basis to fight fires by a federal agency until 1971, when the Bureau of Land Management put an all-female firefighting crew to work in Alaska. The Forest Service reluctantly followed suit in the continental United States, at first fielding all-women crews, then integrating women into existing firefighting teams. The agency debate about placing women in a dangerous occupation foreshadowed the later national debate about women in the military; both centered on whether women had the strength and temperament for traditional male jobs.[429]

By 1978, women had joined hotshot crews and helitack units, in which firefighters rappel from helicopters. The following year, Deanne Shulman, a seasonal firefighter since 1974 who had served on a hotshot crew and a helitack unit, applied for and was accepted into the smokejumpers program at McCall. When Shulman reported for training, she was told that she did not meet the minimum weight threshold and was immediately dismissed. As she packed to leave, she learned from some sympathetic male jumpers that, over the years, several men who were underweight had not been dismissed. Allen "Mouse" Owen, a four-foot-eleven, 120-pound Vietnam War vet who had received congressional waivers on the height and weight requirements and had been with the smokejumpers for ten years, contacted her and encouraged her to fight for her rights.[430]

Shulman did not dispute the legality of her termination but argued that the weight requirement had been waived for others and that she should receive equal treatment. When her initial complaint to the forest supervisor proved unsatisfactory, she filed a formal Equal Employment Opportunity complaint. The Forest Service, faced with unwanted media scrutiny over the dismissal, reconsidered and offered her another chance as long as she met the minimum weight when she reported, which she did. Shulman completed the training in 1981 to become the first

female smokejumper in the United States. Other women soon followed, and another closed door was permanently opened.

Other doors had begun to open as well. The Forest Service appointed its first woman district ranger, Wendy Milner Herrett, in 1979. Herrett had started her career as a landscape architect at Region 6 headquarters in Portland, Oregon. As district ranger, she oversaw 346,000 acres on the Blanco Ranger District of the White River National Forest in Colorado.[431] Her appointment foreshadowed another change: unlike other district rangers, she was neither a forester nor an engineer.

The Consent Decree

Forest Service leadership did not formally address the problem of discrimination against women and minorities in the workplace until a lawsuit in 1973 forced them to do so. At the Forest Service experiment station in Berkeley, Gene Bernardi, a female Forest Service sociologist, applied for a position but the hiring supervisor decided to wait for a male applicant. In 1973, Bernardi sued on the basis of sexual discrimination under Title VII of the Civil Rights Act of 1964, as amended by the Equal Employment Opportunity Act of 1972, and won compensation but not the job. She and several other women then filed a class-action lawsuit over the hiring and promotion of women and minorities in Region 5, which covers all of California.

In 1979, the Forest Service agreed to a consent decree, which the district court approved in 1981. The decree meant the agency had to bring its California workforce into line with that of the state's civilian labor force by having women in more than 43 percent of the jobs in each job series and grade. The Forest Service agreed to monitor progress and enforce the rulings. The

Gordon Rowley, who worked for thirty-four years in fire suppression, recalled, "I often wondered why a woman wanted to fight fire and come down black, grubby, teeth all black, bruised shins, sore muscles. I didn't know what they wanted to do that for. And finally my wife said, 'What do you want to do it for?'" Pictured are (left to right) Jennifer Martynuik, Mara Kendrick, Lori Messenger, and Jeanine Faulkner on the Toiyabe in Nevada in 2000. (USDA Forest Service)

Reagan administration argued that the Bernardi decree represented little more than a hiring quota system, and its opposition delayed the Forest Service's efforts to comply, leading U.S. District Court Judge Samuel Conti to extend its terms until 1991; in 1992, the parties agreed to a new settlement that expired in 1994.

Forced to implement the consent decree or find itself in contempt of court, the Forest Service began to increase the number of women at the GS-11 through GS-13 levels to give them the experience and exposure that would qualify them for higher administrative positions. Aiding its efforts was the implementation of environmental laws, such as the National Forest Management Act, that expanded the agency's responsibilities and required more workers with backgrounds in recreation management, sociology, and other nonforestry disciplines, disciplines that many women had entered because they held more opportunities than did forestry. The rapid promotions of women, however, proved a powerfully divisive issue among employees. Many felt that the consent decree put "accelerated" women in an unfair position, forcing them to succeed or be judged as failures. Some did succeed, to the benefit of the Forest Service, but others did not, and both they and the agency "lost." The shift away from the concept of meritocracy in hiring and promotion practices generated resentment within a few years and created a difficult work atmosphere in Region 5.[432]

Though the Forest Service stepped up the recruiting of women following the consent decree, with so few women in management or in the sciences to serve as mentors or role models, women began seeking ways to connect with one another. The journal *Women in Forestry* (now *Women in Natural Resources*) began publication in 1983 "to provide ideas and information for, from, and about women in the forestry profession."[433] The journal gave women a place to voice their concerns and problems, to learn from one another, and to diminish the isolation they experienced in male-dominated land management agencies.

Professional women entering the Forest Service brought with them a different perspective on the relationship between humans and the environment. A survey conducted in 1990 found that "women in the Forest Service exhibit greater general environmental concern than men" and in particular were more in favor of reducing timber-harvest levels on national forests and designating additional wilderness areas. Another survey found that nontraditional professionals (regardless of gender) held beliefs similar to those of the women in the first survey. Subsequent studies have shown little or no difference in attitudes concerning general environmental issues between men and women, but women exhibited "significantly more concern than men about local or community-based environmental problems." Taken together, the studies suggest that the increase in the number of nontraditional employees had a measurable impact on the attitudes of other employees and was changing the agency's management focus. Forest Service employees' values are now more closely aligned with those of the general public they serve.[434]

Charles "Chip" Cartwright, the agency's first African American district ranger, was urged not to join the Forest Service in 1970 because some in the black community associated agricultural work with slavery. (USDA Forest Service)

Minorities and Cultural Biases

While women made their way into new positions in the agency, African Americans held the fewest jobs of any race at all levels. African Americans had to overcome cultural bias not only in the Forest Service but also within the black community itself. When Charles "Chip" Cartwright considered forestry in the 1960s, agricultural careers carried the stigma of field labor during slavery. Cartwright had been discouraged from studying forestry by his college professors for that reason.[435] But Cartwright's summer job as a Forest Service fire lookout made him want to persevere. After graduating in 1970, he became one of the first African American foresters in the agency and was subsequently the first African American district ranger in 1979 and the first African American forest supervisor in 1988. He took charge of Region 3 (Southwest) in 1994 and was succeeded in 1998 by Ellie Towns, the first African American woman appointed regional forester. Shortly after becoming district ranger in Washington's Okanogan National Forest in 1979, Cartwright began working with black community leaders in nearby Seattle, hoping to attract black youths to enroll in the Young Adult Conservation Corps and forestry schools.[436]

Unlike African Americans, American Indians and Hispanics have long been associated with the Forest Service. Because of the agency's early strategy to hire locals who knew the land and its users best, some of the first rangers in the Southwest came from the local Hispanic population.

In fact, three members of one family were serving as rangers before the 1905 transfer, and four Hispanic rangers were listed at the time of the transfer on the nation's most remote ranger district, the Cuyama District, in what is now the Los Padres National Forest in central coastal California.

But those early hiring practices had long been abandoned, and in the 1980s, Hispanic employees in Region 5 filed a class-action lawsuit. The resolution they reached with the Forest Service in 1992 required the agency to actively recruit, hire, and retain more Hispanics. A second settlement agreement in 2002, like the consent decree of 1979, included further measures to bring the number of Hispanic employees in line with California's workforce, of which Hispanics comprise about thirty percent. As of 2003, Hispanics accounted for about ten percent of the Region 5 payroll.

Paul H. Logan may have been the first black forester to work for the agency.
(Courtesy of Adele Logan Alexander)

The Arrival of the *Ologists*

Implementing the National Environmental Policy Act and the National Forest Management Act created demand for new types of employees, such as wildlife biologists, hydrologists, recreation experts, economists, archaeologists, and sociologists—collectively, *ologists*. Some of these new employees questioned the status quo in land management as well as personnel management. Some knowingly risked their jobs—and in some cases, their personal safety—to speak out publicly against land management practices with which they disagreed. The willingness of some to confront the old-guard foresters and engineers earned them the epithet *combatologists*.

There were several reasons for the differences. Studies conducted in the 1980s found that older foresters who had risen to managerial positions had typically joined the agency between ages nineteen and twenty-four years, an impressionable age, during the agency's heyday. They were so loyal to the agency's mission and methods that they were said to wear green underwear, be green-blooded, or speak the green language. They had been indoctrinated in Forest Service culture and were reluctant to question authority. During the 1980s many older timber managers viewed wildlife management and the other nonforestry sciences as an unwelcome constraint on timber harvesting, and they were not shy about voicing that opinion.[437]

In contrast, the ologists had joined at about age thirty, after attending graduate school. Their graduate studies encouraged loyalty to their professions and emphasized independent research and thinking rather than the conformity and uniformity that had characterized past decision making in the agency. The continued emphasis on timber fostered resentment over the low priority given to the other uses they had been hired to help manage, leading some ologists to question making a long-term commitment to the Forest Service. In addition, female ologists often found it harder to fit in with the male-dominated Forest Service culture and to juggle career and family.[438]

The willingness of combatologists to take on their bosses revived a whistle-blowing tradition in the Forest Service that began with its first chief. Gifford Pinchot had challenged Interior Secretary Richard Ballinger and President William Howard Taft over disputed Alaskan coal leases in 1910 and was fired for insubordination. In the 1910s and 1920s, researcher Raphael Zon argued with Chiefs Graves and Greeley on behalf of an independent research branch and was transferred out of Washington for speaking his mind. Arthur Carhart and Aldo Leopold both resigned from the Forest Service in order to freely advocate for their visions of wilderness. In the 1980s, John Mumma and Jeff DeBonis and other combatologists also wanted to see the Forest Service do what they believed was best for the land and for the public. In doing so, they were carrying out Zon's exhortation: "The success of the Forest Service is based on the encouragement of free expression of new ideas. If forestry is to make progress in the States, the same principle should be recognized even if it calls forth resentment from those who do not want or cannot keep pace with new developments."[439]

Budget Cuts and Backlash

Just as all of those pressures intensified, the Forest Service budget was slashed because of the Balanced Budget and Emergency Deficit Control Act of 1985 (more popularly known as the Gramm-Rudman Act). Aimed at reducing the federal deficit, the act forced the federal government to cut payroll and services. The Forest Service saw a twenty-five-percent reduction in staff. Employees in the traditional forestry positions found that the doors flung open for new scientists and women were now marked "exit" for them. Between 1983 and 1992, jobs in engineering and range management decreased, while employment in nonforestry fields generally increased.[440] Some employees took early retirement, taking their expertise with them.[441] Technology contributed to job losses, too. The introduction of desktop computers, especially the Data General system, in the mid-1980s eliminated the need for typing pools and many of the women who staffed them. In all, between 1980 and 1990, the Forest Service eliminated approximately five thousand positions.

The workforce cuts under Gramm-Rudman prompted a backlash against the consent decree of 1979. In October 1985, African American employees in Region 5 filed a class complaint over

The environmental laws passed in the 1960s and 1970s opened up new career opportunities for women and minorities. Here, scientists monitor fish populations on the Ouachita National Forest in Arkansas. (USDA Forest Service)

their "gross under representation" in the workforce. The Forest Service filed a motion to dismiss the complaint on the basis that it was in conflict with the consent decree; the courts dismissed the complaint in 1991.[442] In 1990, four male employees filed suit to stop the consent decree's implementation. When the courts turned them away, three others joined them in filing another suit, this time claiming reverse discrimination. That, too, was dismissed.[443]

Regional foresters in other regions grew resentful when the women they had recruited and trained for professional and technical positions were reassigned to Region 5 to satisfy the consent decree. The transfers increased the number of women working in that region but did not eliminate harassment and discrimination, and so additional lawsuits were filed in the late 1990s. As part of one settlement agreement, the Forest Service established a monitoring council in 2001 at the regional offices in Vallejo, California, to implement an action plan. Unknown persons vandalized the council's office sign on three occasions, an indication of the continuing animosity.[444]

Two Steps Forward, One Step Back

Although their numbers have increased in forestry, range, and engineering, the categories from which most of the agency's line officers have traditionally been chosen, women have remained underrepresented in those fields.[445] Because of the technical demands of these positions, the Forest Service could not easily promote from within: "You can't change a G-3 clerk into a

District Ranger," one male district ranger noted in 1984. The real problem was not race or gender, he said, but experience and education, which take years to acquire. The district ranger suggested that efforts to get women and minorities into those positions and into management should begin with recruiting from colleges, a strategy the agency has been pursuing to ensure that the composition of its workforce increasingly resembles that of the American labor force.[446]

To recruit more minorities, the Forest Service created partnerships with American Indian institutions and the historically black "1890s" colleges, the land-grant schools established by Congress in 1890 when southern land-grant schools refused to admit black students. The Department of Agriculture had long had close ties with the originally all-white schools; in 1987, Chief Dale Robertson and Secretary of Agriculture Richard E. Lyng set up programs with minority colleges in an effort to diversify the workforce.

Even as it sought to redress past problems, the agency took a misstep. Robertson recalled a conference in Atlanta at which agency leaders met with black university presidents and their deans of agriculture:

> We had a slide program, which was well done, showing people in agriculture at work. Guess how many minorities were in that show? Zero. We agriculture folks were so proud of that slide program because it was professionally done, and we just got immediate negative reaction: "Not a black face in your slide program. That's an indication you don't have blacks in very many positions of agriculture, you're not even sensitive in putting together a slide program to show the black audience that you have blacks working in the Department." … I have to admit…Agriculture had been neglecting them and that they needed USDA as their partner to strengthen the 1890s schools. A lot of truth to that.[447]

Robertson took the lesson to heart and focused on creating opportunities for the schools and their students through his 1890s schools initiative, which provided millions of dollars to recruit top minority students for summer and permanent employment. Several agencies within Agriculture began funding full scholarships for black students interested in careers in natural resources, as well as for American Indians and Hispanics, as part of the recruiting process.

Recruiting African Americans is one challenge; retaining them is another. Arthur Bryant and Jetie Wilds saw a need both to engage and inform African Americans working for the Forest Service about opportunities within the agency, and to help the Forest Service communicate better with its minority employees and become more sensitive to diversity and multiculturalism. In 1992, they formed the African American Strategy Group to encourage diversity and retention of African Americans within the Forest Service. In meetings and memos, Bryant and Wilds identified opportunities to close the communications gap between African Americans and agency leaders. Chief Robertson, who had already launched the 1890s initiative, budgeted some $25,000 a year to help the organization develop strategies and hold workshops. Similar programs were established with groups representing Hispanics, Asians and Pacific Islanders, and the disabled.[448]

Accounting for less than four percent of the agency workforce, African Americans often faced problems similar to those of women, but more extreme. Though a female employee might find she was the only woman in a field office, she was not the only woman in town. For African Americans, working in some remote locations meant being the only black person in the entire community. Those who came from urban backgrounds could experience culture shock and isolation when transplanted to rural, predominantly white towns, and some left the agency.[449]

Retention has in fact proved difficult. From 1992 through 2000, the Forest Service had an average of only 1,241 African American permanent employees, and 1,227 as of December 2004.[450] The majority of those employees (61 percent) were in Region 8 (the South) and in the Washington office. When Region 5 (California) and the Southern Research Station in Asheville, North Carolina, were included, the percentage rose to seventy-seven, leading to accusations of de facto segregation. African Americans constitute only 3.3 percent of the total Forest Service workforce, compared with 6.1 percent for Hispanics and 3.9 percent for American Indians.[451]

In addition to providing training to eliminate discrimination and harassment in the workplace, the Forest Service launched several programs, such as Work Force 1995: Strength through Diversity, designed to achieve an "ideal" workforce as defined by the Civil Service Reform Act of 1978. On the whole, diversity programs and improved personnel management practices, combined with the introduction of professionals from nontraditional fields, have had an irreversible impact on Forest Service culture. By 2004, roughly one-third of all district rangers and forest supervisors were women.[452] Implementing policies important to women employees, such as maternity leave and flexible work schedules, which did not exist when Bernardi filed suit, have benefited men as well as women. Career training has helped both male and female employees advance and become more responsive managers in a period when the Forest Service has to serve more forest users with fewer agency resources than ever before.

Despite the progress in hiring and retaining a diverse workforce, problems remain and lawsuits continue to be filed. As one Forest Service employee noted in 1984, "Given the Forest Service's traditional values, it's a big step to open up the organization to women and minorities. It'll take time, but we're getting there."[453]

Twenty-plus years later, with the agency's employment practices under continued judicial scrutiny, the agency is still getting there. Nevertheless, the Forest Service of 2005 looks nothing like what Herbert Kaufman observed in the late 1950s. The career of Sally Collins perhaps exemplifies the difference. She received a bachelor's degree in outdoor recreation and worked as a deputy forest supervisor and assistant planner for the Forest Service, and for four years as a wilderness specialist, environmental coordinator, and mineral leasing coordinator for the Bureau of Land Management. She also holds a master's degree in public administration with an emphasis in natural resources management. Collins, who is married to an oceanographer, served as the forest supervisor for the Deschutes National Forest in Oregon for seven years before becoming the associate deputy chief for the National Forest System in April 2000. Her diverse academic and professional background complements the more traditional forestry background of other agency leaders, and in 2001, Chief Dale Bosworth promoted her to associate chief, the second-highest position in the Forest Service.

Chapter Eight

TRADITIONAL FORESTRY "HITS THE WALL"

Anti-logging protest, Pacific Southwest Region (San Francisco), early 1970s. (USDA Forest Service)

A T MIDCENTURY, MANY FOREST SERVICE LEADERS BELIEVED THAT SCIENCE AND RATIONAL PLANNING HAD BROUGHT SUCCESS IN DEALING WITH CONSERVATION'S GRAVEST PROBLEMS. THE AGENCY HAD DISPELLED THE LONG-STANDING FEAR OF A TIMBER FAMINE. *TIMBER RESOURCES FOR AMERICA'S FUTURE*, PUBLISHED IN 1958 BY THE FOREST SERVICE, SHOWED THAT FOR THE FIRST TIME, TIMBER GROWTH ON ALL LANDS—PUBLIC AND PRIVATE—WAS EXCEEDING THE ANNUAL CUT. THE AMOUNT OF ACREAGE LOST EACH YEAR TO FIRE HAD STEADILY DROPPED SINCE THE 1930S. IN THE 1960S, THE AVERAGE ANNUAL BURN WAS 4.6 MILLION ACRES, DOWN FROM A HIGH OF 39.1 MILLION ACRES IN THE 1930S. REDUCTIONS IN SHEEP AND CATTLE GRAZING ON PUBLIC LANDS IN THE 1940S HAD HELPED REHABILITATE EXHAUSTED RANGE BY THE 1950S AND ALLOWED FOR INCREASES IN THE NUMBER OF ANIMALS GRAZING IN THE FOLLOWING DECADE.[454]

An optimistic and popular Forest Service entered the 1960s confident. The Multiple Use–Sustained Yield Act of 1960 had put its mission into law, removing any ambiguity regarding the purposes for which the Forest Service should be managing land. The accomplishments of the agency beginning with the New Deal programs bolstered the government's faith in its ability to solve problems. The application of science and technology for the betterment of humankind made the agency popular with much of the public until the early 1970s. The agency's continued emphasis on timber management, however, left it blind to other considerations, such as the impact of chemicals on human and animal populations and the public's changing values. Its fervent faith allowed little room for dissent from within or criticism from outside. Moreover, what the federal government assumed was better for humanity was not necessarily true for the environment, as Rachel Carson had demonstrated.

The Herbicide Wars

In limited use before World War II, chemical pesticide usage in the national forests accelerated in 1947, when Congress passed the Forest Pest Control Act, which charged the Forest Service with preventing, controlling, or eradicating destructive pests on both private and public forests. Industrial foresters and the Forest Service considered insecticides necessary to protect timber and range animals from harmful insects. Herbicides provided an efficient way to foster regeneration of economically desirable trees by killing undesirable trees, maintaining fuel breaks, and destroying noxious weeds on rangeland. Over the years, the Forest Service has battled diseases and insects on public and private lands that threaten healthy ecosystems and even entire species, such as oak and elm trees, and its pesticide programs are of on-going significance to State and Private Forestry.

The agency's confidence in science to effectively handle land management problems and to control outcomes reached its peak during the Vietnam War era. Although Rachel Carson's *Silent Spring* had inspired further studies that showed how insect populations adapted to the chemicals and how pesticides killed beneficial parasites and predators along with the targeted insects, and although many entomologists said that "100 percent control or eradication of an insect was neither necessary nor practical to prevent economic loss," the Forest Service continued its use of chemicals.[455] Many of the agency's own biologists opposed the use of DDT, and the Environmental Protection Agency (EPA) accused the Forest Service of conducting inadequate research on the impact of spraying.

After EPA banned DDT in 1972, the Forest Service turned to other sprays—Malathion, Zectran, sevin 4 oil, and Orthene—not specifically banned by the federal government. When those did not prove as effective as DDT, the Forest Service received permission to use it on an emergency basis to combat an infestation of Douglas-fir tussock moth on more than 250,000 acres of federal forests in the Pacific Northwest in summer 1974. Condemnation from numerous local and national groups followed, and lawsuits and protests against the Forest Service proliferated. Leading timber industry companies including Georgia-Pacific and Weyerhaeuser, and the chemical's manufacturer, Dow Chemical, advocated its continued use, however, and agency leaders steadfastly denied its harmful effects. The 1974 application marked the last large-scale use of DDT in the United States.[456]

One herbicide used by the Forest Service, 2,4,5-T, had been developed by the U.S. Army in World War II and then released for domestic use after the war as a weed and brush killer. The Forest Service began using 2,4,5-T on American hardwoods to clear weeds from around shade-intolerant softwood stands. During the Vietnam War, the military sprayed a mixture of 2,4,5-T and 2,4-D, called Agent Orange, to defoliate the hardwood jungle canopy in Vietnam and deny the enemy safe haven. The military used twenty-seven times more herbicide per area unit than was used stateside for weed control. By 1966, studies had revealed that the herbicides 2,4,5-T and Silvex contained TCDD, a highly toxic dioxin which caused certain birth defects in laboratory animals and was suspected of causing illnesses, birth defects, and miscarriages in humans. The federal government soon imposed limited restrictions, such as banning their use around the home and on food crops intended for human consumption, but allowed its continued use by public and private foresters.[457]

Though antiwar protesters succeeded in getting the military to stop using Agent Orange in Vietnam in 1970, aerial spraying of toxic herbicides in national forests continued. The Forest Service was operating under the assumption that a chemical registered for use with the federal

government did not have any significant adverse effects on the environment. The agency needed to conform to the new environmental laws, including the Federal Environmental Pesticide Control Act of 1972, which gave EPA jurisdiction over pesticides. Furthermore, the Forest Service was not conducting any risk analysis on the health effects of chemicals, nor did it fully consider alternatives, such as brush removal or hand spraying.[458]

Debate over the continued use of 2,4,5-T and Silvex, another pesticide containing dioxin, quickly became a national issue as studies provided conflicting results regarding the toxicity of dioxin. Investigations during the 1970s increasingly pointed to the dangers resulting from the use of pesticides containing dioxin. Meanwhile, Vietnam War veterans exposed to the agents began reporting physical, reproductive, and neurological troubles and demanded treatment. EPA-sponsored studies conducted near Alsea, Oregon, showed significantly higher percentages of miscarriages following the spraying of 2,4,5-T and Silvex. On April 27, 1978, Assistant Agriculture Secretary A. Rupert Cutler announced that until EPA finished its latest study, he would oversee all Forest Service decisions to use the sprays.

That same day, Forest Service Chief John McGuire issued a directive authorizing herbicide use only after all other alternatives had been considered. His failure to ban the chemicals sparked the "herbicide wars." Although the EPA studies were rightfully criticized as flawed, environmental groups and others concerned about the continued use of herbicides on public lands greeted nearly every announcement of an upcoming aerial spraying project with protest until in early 1983, EPA issued its final decision to stop the use of herbicides containing dioxin.[459]

The Forest Service was not alone in its use of dioxin-based herbicides. Private timber companies used a popular mixture of 2,4,5-T and fuel oil to control hardwood brush so that pine could be regenerated by natural reseeding. Spraying operations like this one in West Virginia in 1961 had an impact on wildlife and fish populations, as well as human populations. (USDA Soil Conservation Service – Forest History Society)

After DDT was banned in 1974, populations of the brown pelican off the Gulf Coast and the peregrine falcon, along with the bald eagle, began recovering. Researchers found that DDT and other pesticides interrupted the reproduction cycle of several species and put them at risk for extinction. (Tom Iraci – USDA Forest Service)

In Regions 5 and 6 (covering California, and Washington and Oregon), where debate over pesticides raged loudest, the Forest Service and the Bureau of Land Management continued to use other herbicides until March 1984. A judicial ruling in Oregon at that time stated that a government body that used herbicides must fully consider potential human health problems associated with its programs, and that all potential risks associated with their use must be incorporated in the planning process under the National Environmental Policy Act.[460] The use of herbicides on federal lands in Oregon was immediately suspended. Shortly thereafter, Regional Forester Zane Grey Smith, Jr., also issued a moratorium in California on herbicides. Both bans lasted until 1991.

In the meantime, the agency was assessing alternatives to aerial application of herbicides and testing new herbicides, and went about the difficult task of preparing legally defensible environmental impact statements. Herbicide use for timber management resumed in 1991, with spraying conducted on small areas, typically between 3,000 and 20,000 acres. (By comparison, by the late 1970s, herbicide use for noxious weed control varied between twenty-five thousand and sixty thousand acres.[461])

The herbicide wars forced the Forest Service to reconsider its approach to pest control. Cost-effectiveness ceased to be the sole criterion in deciding to use chemicals. Herbicide use is now one of several weapons in the Forest Service arsenal, along with manual and mechanical cutting and controlled burns, as part of its integrated pest management approach. Where pesticides are used, they are applied only "under very exacting conditions and in a carefully supervised manner."[462] The Forest Service considers aerial application only when it offers significant advantages over other options, and not simply because it is the least expensive or most expedient.

The Forest Service in Vietnam

Some of the Forest Service's overseas activities during the Cold War, in particular the Vietnam War, have come to light only recently with the declassification of secret files and the willingness of those involved to share their experiences. Though the full extent of Forest Service involvement is not yet known, the agency "conducted important programs in support of both civilian and military interests in forest management, fire control and employment, and defoliation." Some Forest Service activities in Vietnam provided cover for Central Intelligence Agency operations during the war. In other cases, Forest Service personnel left to work directly for the CIA.[463] Agency employees were involved in tactical, strategic, and logistical decisions, and in activities ranging from the mundane to the mystifying.

The CIA began hiring for its paramilitary operations in the early years of the Cold War and conducted operations in Laos, Thailand, Vietnam, and other Asian countries as well as in Guatemala, Cuba, and Honduras for the next three decades. The quasi-military culture of the Forest Service made its workers attractive to CIA recruiters. Smokejumpers were especially sought after because they already had training in parachuting and air delivery techniques in rough terrain, and they were fit and adventurous. Several of them quickly found work with the CIA as "cargo kickers," men who pushed supplies out of cargo planes, just as they had pushed supplies out of planes to firefighters. Smokejumpers liked working for the CIA because they could jump fires in the United States during the summer and train foreign jumpers or fly overseas missions the rest of the year.[464]

After the Korean War began in 1950, smokejumper Fred Brauer recalled, a CIA recruiter visited the smokejumper base in Montana and asked him to recommend men for work. Brauer's family obligations prevented him from joining, but two of his unmarried friends signed up. The men had steady work for the next three years, operating out of Taiwan and Okinawa to help deliver agents and supplies by air behind communist lines in Korea and the People's Republic of China. Several former smokejumpers went on to have long and colorful careers with the CIA. Smokejumpers also trained CIA officers and foreign agents in rough-terrain parachuting techniques at their Montana base.[465]

When the Korean War ended in 1953, employment opportunities dropped off, only to pick up again in 1956 as the CIA became more deeply involved in fighting China's takeover of Tibet. Secret operations were launched from numerous countries and lasted until the early 1970s. The CIA used former smokejumpers throughout Southeast Asia for many of those operations. In early 1961, all but two of the smokejumpers on the CIA payroll were transferred to Latin America as the United States prepared for the Bay of Pigs operation. Four or five of the men Brauer had personally recommended for hiring were involved in the failed invasion of Cuba.[466]

From 1959 to 1965, the CIA secretly trained Tibetan soldiers on former Forest Service land at Camp Hale in the heart of the White River National Forest. Camp Hale remained under military control after the war and was made available to the CIA. The remote location was ideal for this secret work. The Tibetans trained with live ammunition, but using white phosphorus rounds instead of blanks on the firing range nearly compromised operations during the dry summer months in 1960, when a bullet ricocheted into the trees and set the hillside on fire. The CIA staff could not cope with the fire themselves but were hesitant to call in the Forest Service and expose the top-secret project. Staff hurried the Tibetans back to their quarters before the fire trucks arrived. The firefighters were told not to ask too many questions, although during the week it took to bring the fire under control, they did

Herbicides and Combating Illegal Drugs

One use for herbicides is in combating illegal plants that are grown on national forests. When cannabis, from which marijuana is produced, first arose as a management problem on national forests in the 1970s, the plant was grown in small plots by a few individuals. By the early 1980s, it had become a big business that relied on heavily armed migrant workers to raise the crops and patrol the fields. Federal agencies sprayed herbicides on plantations as part of its eradication program until those efforts were legally challenged in 1983 because of concern that smoking marijuana that had been sprayed with the herbicide paraquat might be lethal. The Forest Service and other land management agencies agreed to conform to federal regulations before conducting any spraying operations. Plantations are typically located in remote backcountry areas and hidden under the forest canopy, making them virtually impossible to spray with herbicides to destroy the plants without affecting the environment in some way. Since the tightening of U.S. borders following the terrorist attacks on September 11, 2001, domestic production of cannabis has risen dramatically on public lands.

Source: Various news reports, and "Law Enforcement: Marijuana Eradication" file in U.S. Forest Service History Collection, Forest History Society, Durham, NC.

PLEASE!

SMOKEY

Only you can prevent a forest

The success of the Operation Ranch Hand defoliation missions in Vietnam led team members to modify Smokey Bear's motto on a Forest Service poster. (Courtesy of the Ranch Hand Association Vietnam Collection, The Vietnam Archive, Texas Tech University)

notice the "strange inscriptions" in a foreign language that had been carved into some rocks.[467]

The CIA also owned private airlines, such as Air America, and needed experienced pilots, cargo handlers, and airplane maintenance men to staff them. Again, the agency turned to Forest Service smokejumpers. Air America moved equipment and personnel around Southeast Asia when using U.S. military aircraft was undesirable. "Undesirable" sometimes meant flying where the U.S. military was not supposed to be, like Cambodia or Laos, but more often than not, it meant dangerous—maneuvering slow-moving aircraft at low altitudes to "bird dog" enemy positions for jet fighters, conducting aerial spraying, or dropping cargo to troops in the field while taking enemy fire. Air America pilots also ferried Forest Service personnel and others working for the U.S. Agency for International Development.[468]

From 1965 to 1967, Forest Service researchers from its Montana and California fire research laboratories were in Vietnam advising on secret projects. Operation Ranch Hand, which became operational in 1962 and ended in 1971, used Agent Orange and other defoliants to open up the hardwood jungle canopy to expose enemy

movements. Poor initial test results were no deterrent. The Military Assistance Command–Vietnam ordered additional spraying using formulas with more dioxin. The military command then expanded its list of targets to include food crops, both to starve the enemy and to drive the South Vietnamese off the land and into internment camps.[469]

Forest Service fire researchers also worked on the Sherwood Forest and Pink Rose operations, which involved chemically defoliating the jungle to create dry fuel and then dropping incendiary weapons, such as magnesium firebombs, to start a firestorm. Sherwood Forest began in January 1965 with the intensive bombing of Boi Loi Woods, a dense forest twenty-six miles northwest of Saigon that served as an enemy stronghold. Airplanes spent two days dropping eight hundred tons of bombs before the spray planes began dispensing 78,800 gallons of herbicide over the next twenty-nine days. Forty days later, after the foliage had fallen and the vegetation had dried, bombers dropped diesel fuel and incendiaries. The rising heat from the fires, however, triggered a rainstorm over the burning forest that doused the flames. The defoliant operation opened up the canopy as hoped, but only temporarily. The quick return of enemy forces to the area soon

Jay Cravens, who visited all forty-four provinces of Vietnam while serving there with the Forest Service, recalled that everywhere he went, the country reeked of herbicide. He lived in Green Beret compounds while visiting sawmill operations, and always carried a weapon. This photograph was taken while he waited for a jeep to take him to the Tra Bong sawmill. (Courtesy of Jay Cravens)

thereafter indicated that chemical agents alone would not deny the enemy permanent use of the Boi Loi Woods.[470]

Pink Rose, which began in May 1966 and ended a year later, was another attempt to set fire to the jungle. Planners decided to defoliate the chosen areas three times over the course of a year before igniting the dried vegetation with incendiary bombs.[471] The military had high hopes for Pink Rose and even sent up a planeload of journalists to watch the burn experiment. Results, however, were similar to Sherwood Forest—the heat created rain clouds that extinguished the fires. The military discontinued the firestorm experiments, which one government official later admitted was a "nutty" idea to begin with.[472] Defoliation operations to expose communication and travel routes used by the Viet Cong continued in South Vietnam and then expanded into Laos in December 1965 and into North Vietnam in summer 1966.[473]

In January 1967, as fighting in Vietnam escalated, the Forest Service lent a seven-man team of foresters to the U.S. Agency for International Development to help conduct forestry operations in South Vietnam. The team was dispatched after Chief Cliff and other forestry and logging experts visited Vietnam in 1966 at the request of the secretary of Agriculture to study the lumber supply situation. Most of the lumber used by the military was being shipped from the United States, which created logistical problems. After examining the situation, Cliff agreed to supply Forest Service personnel to help increase local production of lumber and plywood.[474] Led by Jay H. Cravens, whom Cliff personally selected, the Forest Service team advised local populations on logging, milling, and reforestation efforts. The locals would become economically self-sufficient, planners hoped, and not side with the Viet Cong. No one publicly questioned the logic of trying to conduct logging operations in a war zone.

In the end, the ill-conceived forestry program helped undermine American efforts. The Viet Cong demanded bribes from loggers and infiltrated operations. The best sawmill operator that Cravens trained turned out to be the leader of the local Viet Cong unit.[475] Although the United States military provided logistical support and military protection for the foresters as they flew around South Vietnam to advise on logging operations and set up sawmills, the military also continued its defoliation and bombing missions, often near the proposed logging operations. Damage to vegetable, fruit, and rubber tree farms angered farmers, and shrapnel in tree trunks wreaked havoc with saw blades at the lumber mills.[476] The mission was a political and economic failure.

Cravens meanwhile advised top military commanders to stop using the defoliants because of their ineffectiveness and long-term effects on the forests, to no avail.[477] Though Ranch Hand missions cleared jungle canopy, bamboo quickly sprang up and provided thick cover, as Cravens had warned. Local populations and soldiers alike suffered from dioxin poisoning from handling the chemicals, being in or near spray zones, or conducting military operations in defoliated areas.[478]

Earth Day

After sixteen months in Vietnam, Jay Cravens returned home in May 1968 to find a changed United States. Before he left, Rachel Carson's *Silent Spring* and the reissue of Aldo Leopold's land ethic essay, which advocated ecologically balanced land management policies, had begun to awaken the general public to environmental problems. By the time he returned, the antiwar movement was powerful, the environmental movement was rapidly growing, and the civil rights movement was turning violent. President Lyndon Johnson had announced in March 1968 that he would not seek reelection later that year. Civil rights leader Martin Luther King, Jr., had been slain in April, and Democratic presidential hopeful Senator Robert Kennedy was running hard in a campaign that would end in his death in June.[479]

When Richard Nixon became president the following year, he recognized the political importance of the environmental movement. He signed the National Environmental Policy Act into law on January 1, 1970. By executive order, he created the Environmental Protection Agency in 1971 to regulate environmental laws passed by Congress. Another major initiative, however, was defeated. When Nixon proposed a Department of Natural Resources, which would have resulted in the dismantling of the Forest Service, Congress resisted and the idea died a quiet death. Similar attempts to reorganize land management agencies under President Jimmy Carter a half-dozen years later met the same fate.[480]

The year 1970 also saw the first celebration of Earth Day, the culmination of seven years of work by Wisconsin Democratic Senator Gaylord Nelson. Troubled by the growing number of environmental problems, Nelson tapped the energy and organizational efforts of the antiwar movement to plan a grassroots protest against the degradation of the environment. About twenty million people turned out across the nation. The Nixon administration made token efforts to show interest in what it perceived to be a left-wing, anti-administration event, but shortly after Earth Day, it returned its attention to the Vietnam War.[481]

The Forest Service, however, welcomed Earth Day. In 1970, Chief Ed Cliff circulated a memo to all Forest Service employees that read, in part, "Our programs are out of balance to meet public needs for the environmental 1970s and we are receiving criticism from all sides…. The Forest Service is seeking a balanced program with full concern for the quality of the environment."[482] In September 1971 the Forest Service introduced Woodsy Owl, its antipollution "spokesman," who urged the public to "Give a hoot! Don't pollute!" In 1997, the Forest Service updated his message to "Lend a Hand, Care for the Land."

While introducing Woodsy Owl, the Forest Service faced continuing controversy over clear-cutting on the Bitterroot National Forest and debate over the 1972 Roadless Area Review and Evaluation, which had not included eastern national forests. And the Forest Service in 1973 was instructed that timber programs would receive the greatest share of its budget.[483] Meanwhile, the Vietnam War dragged on and revelations about the Pentagon Papers, My Lai massacre, and the secret invasion of Cambodia, and Watergate eroded public trust of governmental leaders. Forest Service leaders were no exception.

The Nixon administration's attempts to gut the Environmental Protection Agency energized environmentalists. The Watergate scandal and energy shortage crisis enveloped Nixon's presidency in 1973 and left him unable to stop Congress from passing the Endangered Species Act of 1973. The federal courts ruled against the administration in the Monongahela clearcutting case that same year, setting the stage for passage of the National Forest Management Act in 1976. Both laws portended great changes for the Forest Service.[484]

In 1971, Woodsy Owl joined Smokey Bear in the agency's public affairs program to urge Americans to fight pollution. The change in land management policies in the 1990s led Smokey to modify his original message from preventing forest fires to preventing wild fires, while Woodsy broadened his antipollution message to "Lend a Hand, Care for the Land." (USDA Forest Service – Forest History Society)

When John McGuire became chief in 1972, his colleagues gave him this cartoon drawing. Contending with the issues "sharks" will always be part of the job of Forest Service chief. (Courtesy of Marjory McGuire)

The Forest Planning Process

Until passage of the National Environmental Policy Act, there had been no legal or regulatory requirements to involve the general public in decisions regarding national forest and resource management plans. The appeals process established by the Forest Service in 1906 had been intended to garner the support and cooperation of neighbors and users of the national forests. Most appeals were handled on a local basis between the permit holder and the local ranger or the next higher official. The *Light* and *Grimaud* grazing cases of 1911 were notable exceptions: the Forest Service had wanted Supreme Court rulings to help establish legal precedent for its grazing policy. Public involvement in planning remained minimal until the 1960s.

Besides the "big five" land management laws—Multiple Use–Sustained Yield Act (MUSY), the National Environmental Policy Act (NEPA), the Endangered Species Act (ESA), the National

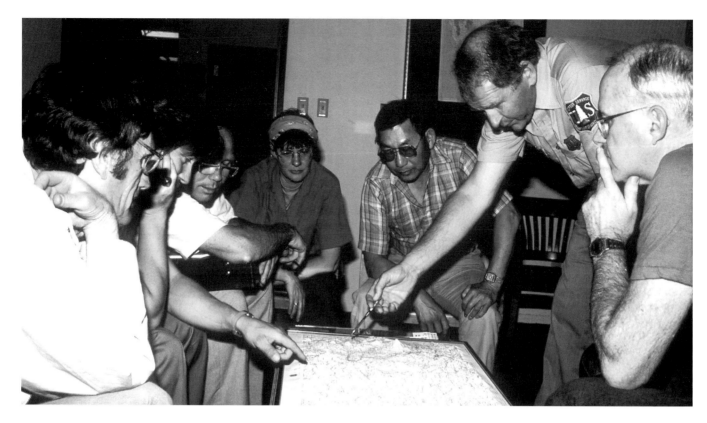

Forest Management Act (NFMA), and the Wilderness Act—nearly seventy other laws govern land management. Plans must conform to those laws, the application of which can create conflict on the ground, and the Forest Service is required to allocate funds to carry out those plans. Congress and the executive branch (through the Office of Management and Budget) set timber production levels and allocate funds for that activity but have consistently failed to fully fund nontimber activities.

NEPA requires federal agencies to disclose "major federal actions affecting the quality of the human environment" and to seek public review on these actions prior to final decisions. NFMA requires predecisional public participation in the development of forest plans. Both laws significantly changed the level and intensity of public involvement in the planning of land management projects. Before the 1950s, only those people or groups directly affected by a project expressed an interest in agency decisions.

The Forest Service issued NFMA regulations in 1979 for operating under the new legislation. For the first time, the agency worked closely with the public in drafting forest plans. Members of the "public"—occasional visitors, backcountry hikers, timber and livestock industry leaders, workers who made their living in the forests, environmental advocates, local communities dependent on their twenty-five-percent share of the receipts from timber sales—all wanted their voices heard.

The National Forest Management Act of 1976 required public input during the forest planning process, as seen here at one such meeting in 1989. Preparing forest management plans can take more than ten years in some cases because of the number and length of legal challenges to management plans.
(USDA Forest Service)

Anti-logging activists protested a timber sale on Mount Hood in 2000 by surrounding the regional office in Portland, Oregon. One protester camped on a ledge above the building entrance for eleven days before being arrested. Protests were commonplace in the Pacific Northwest in the late 1980s as the spotted owl controversy raged. (Tom Iraci – USDA Forest Service)

Immediately following the implementation of regulations, the Forest Service trained its employees on how to obtain and analyze public input on proposals. To avoid appeals, the agency built stronger scientific and economic justifications for its decisions and became more aware of the potential weak points of proposed projects. Many Forest Service officials hoped these measures would reduce litigation, but the new process could not resolve the fundamental differences about the competing uses of the national forests among stakeholder groups.

During the 1980s, conflicts shifted from disputes over the nature and timing of specific projects on national forests to whether some projects should move forward at all. Regulations did not require participation and comments in the planning and design of projects, making it possible to file appeals on issues not raised during the public review and comment periods. The long legal reach of national environmental groups meant that formerly local decisions based on local needs had suddenly become, in effect, national decisions. Project planners and agency environmental specialists had to reanalyze data or develop new information, thus significantly delaying projects, often by two or more years.

The appeals often resulted in "legalistic documents that required a large amount of staff time to analyze, review, and develop decisions for each appeal on each forest plan." Some decisions led to changes in process requirements that other forests then needed to implement. In California, Oregon, and Washington, the source of most Forest Service timber sales in the 1970s and 1980s, changes in process requirements resulted in forest plans that took ten to fifteen years to complete.[485] Forest Service leaders up and down the line became concerned about producing plans that did not comply with the law. The situation became known as "analysis paralysis."

Failure to move ahead with logging or other activities in spite of legal concerns could create other problems. If an injunction halted or prevented cutting, the contract holder could hold the Forest Service liable for third-party damages. Private landholders with lands adjacent to national forest lands could sue over the agency's failure to prevent disease or insect outbreaks from spreading. The buildup of fuel loads in the forests could increase the risk of catastrophic wild fire. Congressional representatives also pressed the Forest Service to continue activities so that their constituents could keep their forest-dependent jobs.[486]

The team effort required by the new legislation to put together a forest plan brought foresters into working relationships with their new colleagues—the sociologists, hydrologists, and other ologists. Team leaders, focused on meeting the allowable sale quantities (ASQs) to fulfill their assignments, seemed to ignore recommendations from ologists, who made public their criticisms of the Forest Service's timber management program. Nonetheless, the agency continued to strive to harvest all the timber called for in the national budgets.

Another Sagebrush Rebellion

By the 1980 presidential election, a combination of high unemployment, interest, and inflation rates had been a drag on the economy through much of the 1970s. During the campaign, Ronald Reagan, the Republican nominee and former governor of California, declared himself a supporter of another sagebrush rebellion—the western movement for commodity development and the "wise use" of natural resources who favored easing environmental regulations and devolving federal powers of land management to the states. His presidential administration also explored selling public land. Transferring federal lands to state control meant not having to contend with the new federal regulations for resource extraction.

These rebels were political descendents of westerners who had tried to overturn the forest reserves in the 1890s, who gathered at the public lands convention in Denver in 1907 to oppose the policies of Theodore Roosevelt and Gifford Pinchot, who had found support for creating the Grazing Service in the 1930s, and who advocated timber harvest increases and public works dams in the 1950s during the Eisenhower administration. The Reagan-era rebels contended that federal policies left them living under a "colonial" system controlled by easterners ignorant of conditions and concerns in the West.[487]

The rebels turned to an amenable President Reagan and Secretary of Interior James Watt for help. Watt came to Washington with a reputation as an anticonservationist from the Colorado-based Mountain States Legal

---

International Forestry and International Programs

The environmental movement had an impact on the agency's involvement in international forestry. Tropical forestry research, however, predates the Forest Service itself. When the United States took charge of the Philippines and Puerto Rico following the Spanish-American War in 1898, the military asked the Division of Forestry for aid in conducting forestry operations. The Forest Products Laboratory took up tropical wood research shortly after it opened in 1910, and in 1940, the Forest Service established the Tropical Research Experiment Station in Puerto Rico. Until World War II, however, U.S. foresters lacked specific authority to work in overseas projects.

When the U.S. government began foreign aid projects after the war, the Forest Service became involved in international forestry programs. The Agency for International Development and the United Nations Food and Agricultural Organization recruited individual foresters in the 1950s and 1960s. In 1958, the Forest Service launched International Forestry but it focused mainly on research and relied on other agencies for most of its funding. Concern over the rapid rate of tropical deforestation in the 1980s led to the creation of the Forestry Support Program and a flurry of publications. The Global Climate Change Prevention Act of 1990 directed the secretary of Agriculture to make the new Office of International Forestry one of four main branches of the Forest Service. After an agency reorganization and budget cuts in 1997, International Forestry was renamed the Office of International Programs. It has three main units: Technical Cooperation, Policy, and the Disaster Assistance Support Program.

Source: Gerald W. Williams, The USDA Forest Service—The First Century *(2000), 138–41; and Terry West, "USDA Forest Service Involvement in Post World War II International Forestry," in* Changing Tropical Forests: Historical Perspectives on Today's Challenges in Central & South America *(1992), 277–91.*

In order to lessen the impact on the environment, the Forest Service began using helicopters in the 1970s to conduct logging operations in remote or difficult terrain. (Jim Hughes – USDA Forest Service)

Foundation, a sagebrush rebellion think tank; by October 1983, his confrontational style had worn thin and the administration forced him from office. By this time, the rebellion as a movement had largely run its course but its ideas remained popular with the administration. Watt's successors at Interior, William Clark (1983–1985) and Donald Hodel (1985–1989), supported the rebellion's development agenda but went about their business more quietly.[488]

The Reagan administration wanted to jumpstart the home construction industry to help end the economic recession. Assistant Secretary of Agriculture for Natural Resources John Crowell asked the Forest Service to increase timber sales by thirty percent by 2000 and to nearly double timber-cutting levels to nearly twenty billion board feet by 2030. Senior members of Congress from Oregon and other timber states began attaching amendments to Forest Service budget legislation containing specific timber harvest quotas to keep production high.[489] Increasingly, forest managers went after timber in remote backcountry areas to meet their quotas. Those

levels of production would bring significant environmental and economic costs.[490] Max Peterson, appointed chief in mid-1979, found himself at odds with the Reagan administration's desire for high timber quotas and left the agency.[491]

In another attempt to reduce the environmental impact of logging, in 1980 the agency had the U.S. Navy construct a helistat, in which the static lift of a helium-filled aerostat was augmented by the dynamic lift of four helicopters to achieve even greater lift. The operation proved too expensive and the experiment was abandoned.
(USDA Forest Service)

FORPLAN and Forest Planning

Crowell took his cue about what the Forest Service could achieve in part from the methods and computer models the agency used to develop its forest plans. FORPLAN, the linear-programming computer model for forest planning introduced by the agency in 1979 to project future values and impacts of management, was designed to calculate sustainable levels of timber output, or nondeclining yield, over long periods. Until its implementation, national forest planners used a variety of tools, which made it difficult to coordinate efforts across forests, districts, and regions. FORPLAN's designers recommended testing it on three forests for two years. Instead, the Forest Service, seeking to standardize forest planning across the national forest system, adopted the program universally.[492]

FORPLAN tended to make optimistic forecasts and inflate timber production output assumptions. Reflecting the emphasis of the time, the program was timber focused. Though designed for multiple-use management, it balanced other uses of the land against timber production. The program also scheduled management activities by acre, which failed to take into account how activities in one area might affect ecological functioning in another. By focusing on timber harvest scheduling, FORPLAN tended "to study intensively and solve the wrong problems" and was likely to "intensify rather than solve the problem." The problems were recognized soon after its implementation but difficult to correct.[493] When the Forest Service tried to reorient itself toward ecosystem management in the late 1980s, the program proved inadequate. As FORPLAN designer Norm Johnson noted in 1995, "When you go from measuring sustained yield of outputs to sustaining processes and functions of the ecosystem, you really change your whole orientation and how you think about the problem. FORPLAN was not set up to deal with the latter view of sustainability."[494]

The computer models used since the 1950s had all contributed to what historian Paul Hirt has called the Forest Service's "conspiracy of optimism," its faith that technology could produce greater and greater forest outputs. For three decades, timber harvest models had been telling forest planners that the forests could produce ever-higher timber yields, but the models failed to account for other factors such as budget cuts or environmental conditions. Doing what they had been trained to do and using computer models built around timber, forest planners subordinated other forest uses.[495]

Orville Daniels, who served as forest supervisor on the Lolo and Bitterroot National Forests in the 1970s, remembered how timber-focused the agency was:

> We started off cutting timber for a very specific purpose, which is to get young thrifty stands. And as we got into that and more and more of budget and our program got tied up in the timber harvest, all sorts of things then became dependent upon the timber budget. The status of your forest, the grade that you got, [and] the pay you got as the forest supervisor depended on having a big timber program. The regional foresters' staff was funded from the timber program. Sometimes as much as 60 [to] 70 percent of the whole region's budget was coming out of timber. So having high timber targets, high timber outputs and meeting those was a very, very significant part of your performance evaluation. And, if you could not meet your targets, it was not uncommon to be removed from your job. Now, [when] they removed people from the job for not meeting targets most of the time it was timber, sometimes it was other things. So, you could lose your job. The threat was always there.[496]

Regardless of the reasons, Forest Service planners and agency leaders did not question how higher harvest levels could be reconciled with higher standards for environmental protection.[497] Others in the agency, however, soon did.

Dissension in the Ranks

In 1987, the Forest Service harvested nearly fourteen billion board feet on the national forests, the most ever. As the harvest levels increased in the 1980s, so did the protests. The environmental cold war was turning hot just as the ideological Cold War, which had spawned the drive to get

out the cut, was winding down. By the time the Berlin Wall was dismantled in autumn 1989, the Forest Service was locked in a struggle with environmental advocates, who quickly became "the enemy" to many Forest Service employees. Patricia Woods, a consultant with the Forest Service at that time, recalled,

> We had a big discussion in one of the management policy seminars about environmentalists and there was a certain kind of hostility in the class towards environmentalists. And I was incredulous. I said, "Aren't you environmentalists?" "Oh, no, we're not environmentalists. They sue." I said, "Excuse me? Do you fish, do you hunt, do you camp?" "Yes." I said, "Well, don't you want to have healthy lands out there to do that on?" "Well, yes, but we're not environmentalists, we're conservationists because conservationists use the land." There seemed to be this notion that environmentalists were just like the Park Service, that they were all into preservation. I think there was a misunderstanding, maybe, as to who were environmentalists and what environmentalism was.[498]

Two views of the Forest Service: By the 1980s, environmental activists believed the agency had become too willing to cut timber and too greedy (left), while agency employees felt kicked around and misunderstood by the public. (USDA Forest Service – Forest History Society)

The Forest Service consistently lost court cases over clearcutting and viable populations of wildlife species, and it was increasingly at odds with other agencies like the U.S. Fish and Wildlife Service. Project planners and agency environmental specialists had to produce new information to win Fish and Wildlife Service approval, which significantly delayed projects. The Forest Service, as Chief Dale Robertson (1987–1993) pointed out, became fearful of producing plans that would not withstand judicial scrutiny. In Robertson's words, traditional forestry had "hit the wall."[499]

To get out of the quagmire, Robertson announced that the Forest Service was adopting a different management approach. Through New Perspectives, the agency proposed to bring timber management into line with other forest uses, and to handle the land with greater sensitivity toward ecological systems. This approach implied scaling back on timber harvest levels and other extractive activities and preserving or rehabilitating habitats. It required the entire culture and focus of the agency to change with it.

Change was slow and difficult, however. In 1989, following a national forest supervisors' meeting in Tucson, called the Sunbird Conference, 123 national forest supervisors gave Chief Robertson a memorandum complaining that the allowable sale quantities for most forests were "unrealistically high even with full funding" and contributed to the degradation of land quality. They said that the continued underfunding of noncommodity values (thirty-five percent of the National Forest System budget in the previous twenty years had gone to the timber program versus an average of two to three percent each for recreation, fish and wildlife, and soil and water) left the impression that the Forest Service did not support its new mission statement, "Caring for the land and serving people." The memo listed recommendations for improving management and for reestablishing trust in the agency, and the supervisors pledged to work with the chief to restore the agency to a position of conservation leadership.[500]

The memo was one of many personnel headaches for Chief Robertson. Having deftly recovered from the embarrassing presentation to the presidents of the 1890s colleges in 1987, he was still dealing with the consent decree of 1979 and hiring practices in Region 5. Also in 1989, Jeff DeBonis, a timber sale planner on the Willamette National Forest, founded the Association of Forest Service Employees for Environmental Ethics (AFSEEE) to give voice (sometimes anonymously) to those wishing to work for reform. The small watchdog organization (its membership represented less than five percent of the Forest Service workforce but counted retired agency personnel and representatives of environmental groups among its members) published a newsletter in summer 1989 that questioned the agency's timber policy and its commitment to protecting biological diversity.

Like the forest supervisors who signed the memo to Robertson, AFSEEE supported calls for better ecological management by stressing biological diversity and sustainability instead of only the "short-term political expediency" of timber production. Though it agreed that timber harvesting was appropriate on public lands, the group asked that it not continue to be "at the expense of other resource values or public subsidies."[501] The organization and employees who spoke out were criticized within the agency for violating an unwritten code of taking agency problems public, and by the timber industry for making assertions regarding overcutting that it disputed.[502] It was not uncommon for defenders of the status quo, including industry figures, to seek the dismissal of dissenters working for the Forest Service.[503]

The Forest Service had endured external criticism for years, but internal criticism taken public was something new. Dissent and disagreement began to affect its operations. Robertson not only had to defend his actions to Congress and the public, but now to his own employees. John Mumma, the first wildlife biologist to hold the position of Region 1 Regional Forester, first directed his forest supervisors in Montana to increase their timber sales in 1990, despite their protestations that they should instead lower timber targets on their forests. Some said they would have to violate environmental laws in order to meet the targets. Mumma then reversed his position and—going against the wishes of agency leadership and Congress—told the supervisors that if they had to choose between protecting the environment and meeting congressional timber harvest targets, they should choose the environment.[504]

Some of the forest supervisors then took the unprecedented step of writing a joint memo to Chief Robertson in September 1991, stating that the goal for Region 1 should be lowered from 940 million board feet to no more than 590 million board feet. Someone leaked the memo to the press, placing Robertson in a difficult position. He, too, wanted to lower timber harvests, but could not do so dramatically without political fallout on Capitol Hill. Robertson felt pressured by proindustry members of Congress to meet the goals and hold the rebellious supervisors at the regional and forest level accountable.[505] He promised Congress that Region 1 would in fact deliver at least 664 million board feet and hoped to produce 750 million.[506]

Mumma was also under investigation for improperly purchasing horses for the region's use.[507] Robertson, convinced that he had an uncooperative regional forester on his hands, met with Mumma to get him back in line. When Mumma failed to do as Robertson asked, he offered him the choice of moving to Washington or retiring. Mumma retired and quickly became a martyr in environmental circles. Many viewed the transfer offer as retribution, even though the Forest Service routinely rotated personnel within the Senior Executive Service about every four years. When his resignation made it into the national press, it led to a congressional investigation into the broader problem of possible mistreatment of government whistleblowers.[508] The Forest Service's commitment to its new land management approach, New Perspectives, was now under fire from all sides.

Chapter Nine

A NEW LAND ETHIC
AND ECOSYSTEM MANAGEMENT

A second-growth forest on the west side of Oregon's Cascade Mountains. (Tom Iraci – USDA Forest Service)

TRYING TO MANAGE FOR MULTIPLE USES WHILE COMPLYING WITH MYRIAD ENVIRONMENTAL LAWS AND SEEKING TO SATISFY BOTH CONGRESS AND THE PUBLIC CHALLENGED THE FOREST SERVICE IN THE 1980S. WITH MORE THAN 40,000 EMPLOYEES, IT COULD NOT CHANGE HOW IT MANAGED LAND OVERNIGHT, NOR WAS THE PATH ALWAYS CLEAR, GIVEN THE LIMITS OF SCIENTIFIC KNOWLEDGE. UNDER INTENSE PRESSURE FROM BOTH INTERNAL AND EXTERNAL CRITICS, THE FOREST SERVICE RESHAPED ITS APPROACH TO MULTIPLE-USE MANAGEMENT. IT TURNED TO THE LAND ITSELF FOR CLUES AND REDEFINED THE GREATEST GOOD YET AGAIN.

Origins of Ecosystem Management

Aldo Leopold's posthumous work, *A Sand County Almanac and Sketches Here and There*, was overlooked outside conservation circles when it was published in 1949. When *A Sand County Almanac* was reissued in paperback in 1966, however, the collection of essays, especially "The Land Ethic," catalyzed the environmental movement and influenced the education of a new generation of natural resource scientists.

"The Land Ethic" was the culmination of a lifetime of experience and reflection. Writing in 1947, Leopold combined ideas from three essays composed over the previous fourteen years. From "The Conservation Ethic," written in 1933, he borrowed the discussions of the "ethical sequence" and the ecological interpretation of history. A 1939 essay, "A Biotic View of Land," provided the scientific backbone. "Ecological Conscience," written earlier in 1947, argued for a new guide in land-use decisions: "The practice of conservation must spring from a conviction of what is ethically and esthetically right as well as what is economically expedient. A thing is right only when it tends to preserve the integrity, stability, and beauty of the community, and the community includes the soil, waters, fauna, and flora, as well as the people....If we grant the premise that an ecological conscience is possible and needed, then its first tenet must be this: economic provocation is no longer a satisfactory excuse for unsocial land-use (or, to use somewhat stronger words, for ecological atrocities)."[509]

In "The Land Ethic," Leopold reiterated and expanded those ideas. His land ethic recognized the need for management but proposed an approach different from that of traditional forestry and agriculture. Economics still had its place in management plans because "[a] land ethic of course cannot prevent the alteration, management, and use" of soil, water, plants, and animals, "but it does affirm their right to continued existence, and, at least in spots, their continued existence in a natural state." He made it clear how humans fit into the community: "[A] land ethic changes the role of *Homo sapiens* from conqueror of the land-community to plain member

and citizen of it. It implies respect for his fellow-members, and also respect for the community as such."[510]

Leopold expanded "the greatest number" in that statement by restoring humans to the landscape and asserting that *Homo sapiens* are part of the biotic community. Conservation was not about profit, nor was it about sentiment. The land ethic had to combine the practical (Pinchot and conservation) and the sentimental (Muir and preservation) with the intellectual (Elton and ecology). Balanced thinking and management, in Leopold's mind, contributed to the balance of nature.

Aldo Leopold became a University of Wisconsin professor after leaving the Forest Service. His association with scientists in different fields yielded him a broader understanding of ecology than most game managers had in the 1930s. (Courtesy of the Aldo Leopold Foundation Archives, the University of Wisconsin-Madison Archives, and the Robert McCabe Family Collection)

A year before *A Sand County Almanac* first appeared, the Forest Service began research on managing lands on an ecosystem-wide basis in the Pacific Northwest. Operating in the H. J. Andrews Experimental Forest near Blue River, Oregon, researchers examined the practical problems that arose in converting old-growth forests to young forests. By the 1960s, university and Forest Service researchers were looking at the effect of timber cutting on nutrient cycling and productivity, forest-stream interactions, and other ecosystem-level phenomena in old-growth forests.[511]

The Andrews Ecosystem Research Group realized that previous research efforts had failed to measure the impact of large organic debris, such as standing dead trees (or snags), or branches and leaves in streams that provided nutrients and habitat for fish and other organisms. As the interconnectedness of ecosystems became clearer, perceptions about debris and areas disturbed by cutting, fire, or other interruptions began to change. Land managers had traditionally viewed debris as an impediment to logging and a fire hazard, and as obstructions to fish migrating to spawning grounds. Great sums of money had been spent to remove this "waste." Researchers at Andrews, in contrast, recognized the ecological importance of woody debris in preventing erosion and providing habitat.[512] Former Forest Service research scientist Jerry Franklin summarized the differences in how the researchers looked at the forest: "I remember one presentation that I made about old-growth forest where I made quite a large deal about the amount of spiders and fungi in old-growth forest and how diverse and rich it was. And the timber industry fellow that followed made quite an issue of it. He liked bright, well-lit forests without spiders."[513]

In the 1960s, the Forest Service began conducting similar research at the Hubbard Brook Experimental Forest in New Hampshire. Funded largely by the Forest Service and the National

A researcher takes a measurement at the H. J. Andrews Experimental Forest, Willamette National Forest in Oregon, in 1953. Long-term research projects like those at Andrews have produced important findings on forest and grassland ecosystems that continue to inform policy and decision making. (USDA Forest Service)

Science Foundation, the Hubbard Brook research took a multidisciplinary approach to watershed research that looked at the ecosystem; it provided "the basis for future ecosystem research design and provided extensive empirical data for use in resource management." More than 150 scientists took part in the interagency work at Hubbard Brook and produced more than 450 research articles. They examined the movement of water and the nutrient composition of water, revealing the complex relationship between forest and waterways. Research that linked Hubbard's "small ecosystem with the larger biogeochemical cycles of the earth" showed, for example, that forest cutting in New Hampshire could affect water supplies and water quality in Massachusetts, and "created intellectual links between ecology, geochemistry, forestry, soil science, hydrology, and atmospheric science," which greatly expanded the ecological focus beyond watershed research's traditional biology-only focus.[514] Both this work and the Andrews research contributed to a new management approach initially called New Forestry when it emerged in the 1980s.

Unexpected Outcomes at Mount St. Helens

After lying dormant for more than 120 years, a volcano on the Gifford Pinchot National Forest in Washington State erupted early on May 18. The violent eruption removed some thirteen hundred feet from the summit of Mount St. Helens and sent a pillar of smoke and ash eighty thousand feet in the air. A mudflow of melted snow mixed with ash, dirt, and volcanic debris raced down the mountain, carrying logs and sweeping away bridges before depositing

enough silt in the Columbia River to impede navigation. The blast and the resulting avalanche killed fifty-seven people. Forests of fir, spruce, hemlock, and western red cedar were leveled. It looked like someone had spilled a box of giant matchsticks.[515]

The above damaged nearly 62,000 acres of national forest lands and 89,400 of state and private lands, including 68,000 acres belonging to the Weyerhaeuser Company. On national forest land, ash and pumice were more than eight inches deep. Initial estimates put the total monetary loss to the Forest Service at $134 million, including timber (estimated at one billion board feet), road and bridge damage, and fish, game, and recreation values.[516] Coming during a period of debate over the direction the Forest Service should take, the volcanic eruption on the Gifford Pinchot National Forest seemed

On May 18, 1980, a magnitude 5.1 earthquake shook Mount St. Helens (above). The slopes of Smith Creek valley, east of Mount St. Helens, show trees blown down by the blast. Two U.S. Geological Survey scientists (lower right) give scale. More than four billion board feet of timber, enough to build 150,000 homes, was damaged or destroyed. (Austin Post – US Geological Survey [eruption]; Lyn Topinka – US Geological Survey)

like the Old Man himself had spouted off in an effort to remind his beloved Forest Service that it needed to rethink how it managed the land.[517]

At the urging of scientists from a number of agencies, Congress passed legislation creating the 110,000-acre Mount St. Helens National Volcanic Monument and placing it under the Department of Agriculture. The Forest Service moved quickly to salvage 200 million board feet of blown down and standing dead timber from 10,000 acres before any outbreaks of disease and insect infestations could spread to adjacent forests. Outside the monument's boundaries, the Forest Service and private companies conducted salvage timber operations for the same reasons. Weyerhaeuser salvaged more than 850 million board feet of timber, and then planted 18.4 million seedlings on 45,500 acres. To date, nearly 10 million trees have been planted on more than 14,000 acres of national forest land.[518]

The catastrophe presented an unprecedented opportunity for researchers to test their theories of how a forest regenerated in a severely disturbed area. They expected a long, slow process of waiting for microbes, plants, and animals to migrate or be transplanted from adjacent areas. Instead, scientists, including Jerry Franklin and James A. MacMahon, found that the "reality defied predictions." The land began to recover quickly. The timing of the eruption had proven fortuitous. Nocturnal animals that had been underground during the morning blast survived and almost immediately began mixing volcanic ash into the soil and dispersing seeds as they moved about. Winter snow and ice shielded many organisms, and when the snowcap did melt, intact tree saplings and shrubs resumed growing and amphibians emerged unscathed. Snags and downed trees fulfilled their roles in the ecosystem by providing protective cover, habitat, and nourishment for a variety of organisms.[519]

The eruption altered not only the landscape but how Jerry Franklin and other scientists looked at the landscape: "Even though we knew better, we were still thinking in terms of disturbances as destroying things, as laying waste, as eliminating everything from a place, and it doesn't. And so out of that came the epiphany of the biological legacy [that both dead and green are trees needed to sustain ecosystems]. And of course out of that came the notion…that you could harvest timber in ways that would be closer to emulating natural processes."[520]

Research on the massive disturbed area around the volcano led to a reexamination of traditional forestry practices, such as clearcut and selection cutting, all of which "focused on the regeneration of trees and not the perpetuation of a complex forest ecosystem." Typical logging practice on federal lands in the Pacific Northwest was dispersed-patch clearcutting, in which parcels of twenty-five to forty acres were interspersed with forested areas. Environmental conditions, including temperature, humidity, and wind, in the remaining small forest patches were drastically affected by the surrounding clearcuts. Furthermore, these practices often altered or simplified the connections that occur in natural forests. After Mount St. Helens, researchers began looking at forests on a broader landscape scale and started viewing timber management through the prism of land management. Their findings led to a different management approach, "New Forestry."[521]

Under New Forestry, alternative silvicultural practices would "utilize the concepts of ecosystem complexity, biological legacies, and viable, healthy landscapes to retain ecological values." Leaving "biological legacy" trees would benefit the ecosystem and bring back species that had been forced out of the area. Computer simulations suggested that clearcuts should be placed adjacent to existing cutover areas and larger forest patches retained for several decades longer than under the old system of dispersed clearcuts. Many of the concepts embodied in New Forestry, Franklin admitted, were not new, but the focus on the maintenance of complex ecosystems was.[522] Rethinking timber harvest and management plans held out the potential for supporting a wider array of values. New Forestry, in sum, was the application of Aldo Leopold's land ethic.

The Land Ethic in Practice

Even before Franklin had given it a name, however, New Forestry was being practiced on the Shawnee National Forest in southern Illinois. Small even by eastern forest standards, the Shawnee incorporated 256,000 acres in four ranger districts scattered about the area; when combined with the intermingled private lands, the forest's boundaries totaled 714,000 acres. Most of the national forest lands had been farms purchased in the Great Depression that were then planted with a mix of hardwoods and nonnative loblolly pine. In the mid-1980s, the forest managers lowered the annual timber harvest from 20 million board feet of hardwoods and pines to just 3.4 million board feet of hardwood timber and decided to emphasize the restoration of native ecosystems—the hardwood forest, barrens, and wetlands—and break up the large pine plantations planted by the Civilian Conservation Corps in the 1930s. Some local landowners agreed to participate, which allowed for more expansive restoration efforts.[523]

In the Missouri Ozarks, private landowner Leo Drey had been practicing uneven-aged management on some 150,000 acres adjacent to the Mark Twain National Forest since the 1950s. A major part of his management approach included conducting a continuous forest inventory that generated new data every five years showing the growth rate of each tree larger than five inches in diameter at breast height. (By comparison, the Forest Service conducted inventories every twelve or so years and less thoroughly. In 1999, the agency's Forest Inventory and Analysis program began shifting to annual inventories.) As the Forest Service turned away from uneven-aged management and employed clearcutting on a widespread basis in the 1960s, Drey had gone in the opposite direction: he managed his forest for multiple use "in a conservative, sustainable fashion for a full array of ecological, social, and cultural values," and did so profitably. Within the professional forestry and forestry education communities, though, Drey's work was dismissed as "just an experiment" that would not work.[524]

In the early 1980s, Forest Service personnel took a closer look at Drey's lands and decided to implement uneven-aged management on 116,000 acres, or about 11 percent of the Mark Twain—even though they remained skeptical of its practicality and potential success. In 1988, a private group called Mark Twain Forest Watchers filed an official appeal with the Forest Service, "contending that environmental assessments for proposed timber sales had to be site specific and assess the effects of uneven-aged as well as even-aged management." Not only did they win on the Mark Twain National Forest, but Chief Dale Robertson issued a directive in February 1989 to all regional foresters to undertake site-specific analyses in implementing forest plans. The Mark Twain reported a shift from seventy percent even-aged and less than one percent uneven-aged sale acres in 1988 to twenty percent even-aged and thirty percent uneven-aged sales in 1991.[525]

Environmentalists called the stands on the west side of the Cascade Mountains ancient forests; timber managers called them overmature and decadent. To the public they were known mostly as old-growth forests. (Tom Iraci – USDA Forest Service)

New Perspectives and Ecosystem Management

The success on the Shawnee and Mark Twain National Forests seemed more the exception than the rule. By 1989, Chief Robertson had concluded that forestry had "hit the wall": any plans to cut timber generated controversy, court cases were tying up Forest Service personnel and derailing entire forest plans as well as individual projects, and Congress and the White House were complaining that the Forest Service was not making its timber harvest numbers. To continue cutting on national forests, approaches needed to change.

The Myth of the Chief

When asked what a typical day for a chief was like, Dale Robertson responded:

"Let me tell you what it isn't like—the myth that most field employees [have] about the chief. They probably think the chief is our leader up there, he's sitting at his desk and he's thinking about the big picture and plotting strategy, what's our best future and really thinking things through and spending a lot of time about the future of the Forest Service. The truth of the matter [is that] the chief doesn't have time to do any of that…it's a hectic pace, hectic job, and everybody thinks they own the chief. The Congress thinks they own the chief so you've got five hundred and thirty-five members of Congress who think they can—and they can—demand your time by picking up the phone. The administration thinks they own the chief, and they do….And the interest groups are constantly interacting with you and demanding something, demanding your time. You have all of these, plus the public, which can trigger action that takes the chief's time by calling a congressman or calling the administration ….The system will eat you up….It was a seven-day-a-week job."

Source: Steen, Interview with F. Dale Robertson, *55, 57, 58.*

In 1989, Robertson was summoned before a congressional committee to explain how he was going to handle timber harvest levels and endangered species. In his testimony he laid out the Forest Service's plan to take "a bigger, broader, new perspective of the forest" based on New Forestry. For lack of a catchier label, Robertson called his plan "New Perspectives" and pitched it as an alternative to timber-dominated multiple-use management.[526] Overnight, what had been a pilot program became the management approach for the National Forest System.

New Perspectives, however, received a tepid response in many quarters. Because it had been unveiled by the chief at a congressional hearing and without any agency-wide notification, most Forest Service employees did not know what to make of it. Some considered it a return to common sense and quickly embraced it; others saw the plan as an unwarranted admission that foresters had been wrong. Some seized the opportunity and immediately changed management direction and practices; others appealed to Washington for more specifics and guidance. New Perspectives was about processes, and the Washington office "emphasized principles, broad objectives, and expected results but did not provide specific direction on how land and resources were to be managed." The decentralized structure of the Forest Service, the lack of funding for new projects on many national forests, and the failure to clearly define New Perspectives at the beginning made implementation challenging. Nonetheless, within two years of Robertson's announcement, some 260 New Perspectives projects were under way on the national forests.[527]

The Forest Service encountered equal difficulty explaining New Perspectives to a skeptical public. Robertson was criticized by industry and congressional leaders who opposed New Perspectives because they believed it jeopardized timber harvest levels. Environmental activists and other critics denounced New Perspectives as a Forest Service public relations gimmick. Moreover, the agency could not immediately halt its operations and change its management approach or its forest plans overnight, and the new policy did not ban clearcutting.[528]

In fact, clearcutting remained the dominant silvicultural method until June 1992. The presidential administration of George H. W. Bush had come under intense criticism for its seeming indifference towards the environment, and for its weak presence at the international meeting on the environment, or "Earth Summit," in Rio de Janeiro. Fearful of political fallout in an election year, administration officials asked Robertson to draft a statement on the elimination of clearcutting three days before Bush was scheduled to appear in Rio. Robertson seized the opportunity to make New Perspectives the agency's official policy under a different name and asked President Bush to announce the change in policy in addition to the end of clearcutting as the main silvicultural approach on national forest lands. The same day the president made the announcement in Rio, Robertson announced it in the United States.

With that, "ecosystem management" became official policy, and the agency no longer used clearcutting as a standard practice.[529] Because the agency received no legislative mandate to manage for ecosystems, however, as it did for multiple uses, a president or Congress can ignore it. Ecosystem management nevertheless provided a new way to manage for multiple uses, one that takes into account the values and objectives of the public for the first time.

Ecosystem management drew on existing ideas but represented a fresh approach to forestry. Faced with conflicting expectations from opposing constituencies, a decline in resource quality and quantity, and an uncertain future, the Forest Service had introduced ecosystem management as an alternative to total preservation or total development by focusing on "the maintenance of complex ecosystems and not just the regeneration of trees." It accommodated "ecological values, while allowing for the extraction of commodities."[530] Jerry Franklin's group envisioned the approach as an evolutionary step along the continuum from Pinchot's utilitarian "greatest good" ethic to Leopold's "land ethic" that would protect "[t]he productivity of our land, the diversity of our plant and animal gene pool, and the overall integrity of our forest and stream ecosystems" on both commodity and preservation land.[531]

Only a small percentage of the land could be treated through management activities each year, however, and new approaches take time to implement. In fact, it took several years to implement an ecosystem approach across the National Forest System—years that count as some of the most difficult in Forest Service history. Congress and the public needed to understand

Launched shortly after he became chief in 1987, Chief Dale Robertson's Rise to the Future fisheries program involved stream rehabilitation. Heavy machinery modified to limit the environmental impact, like this backhoe turned into a "spider," was used to place logs in streams to improve fish habitat. (Tom Iraci – USDA Forest Service)

the concept of ecosystem management, and thousands of professionals needed training to meet new objectives. "Thus," observed Forest Service scientist John Fedkiw, "research frequently became the principal route to finding new management approaches. Scientific studies to develop the new data, knowledge, and technology to successfully implement new approaches" took years.[532] Research was a never-ending process because all three elements continued to change.

Although the impact of ecosystem management on timber management took several years to be realized, the approach had an almost immediate impact in reinvigorating wildlife management efforts. The restoration of habitats made possible the reintroduction of species not seen in many North American forests in decades. The Forest Service helped reintroduce the red wolf to national forests in eastern Tennessee and North Carolina; the gray wolf to national forests in the Southwest, the northern Rockies, and the Great Lakes region; and grizzly bears to the Bitterroot Mountains.

Under Robertson's leadership, the Forest Service introduced new fish and wildlife resource programs. The national forests have 128,000 miles of fishing streams and rivers, 12,500 miles of coast and shoreline, and more than 2.2 million acres of lakes, ponds, and reservoirs. Fishing, one of the most popular recreational activities on national forests, received a major boost from the "Rise to the Future!" program. The program brought together governmental agencies, private industry, researchers, and anglers to enhance the production of fish and protection of fish habitats. Other related programs—some carried out with the cooperation of other state and federal agencies—targeted plant communities, waterfowl and wetlands, and migratory birds for conservation and protection.[533]

The Northern Spotted Owl

Ecosystem management offered Chief Robertson and the Forest Service a new way to discuss the harvesting of old-growth timber in the Pacific Northwest. Old-growth ponderosa pine on the east side of the Cascade Mountains and old-growth Douglas-fir on the west side supplied about half the timber harvested from the National Forest System and nearly all the timber harvested from Bureau of Land Management lands. Timber harvest levels on public lands in the region had remained high through the 1970s and 1980s. The area was also the habitat of the northern spotted owl.

The issue began in 1972, when Eric Forsman, a student at Oregon State University, observed that northern spotted owls were most commonly found in old-growth forests of the Pacific Northwest. Studies showed that the owl population was dropping, and by the mid-1980s, the species was under consideration by the U.S. Fish and Wildlife Service for listing as an endangered species. Three interested parties—government, private industry, and environmental activists—soon clashed over spotted owl habitat.[534]

The Forest Service delayed confronting the problem by ordering study after study of owl habitat, hoping either to find a scientific solution to validate timber management objectives or to delay the issue long enough to get the cut out. Had the agency better balanced timber harvesting and habitat protection when the spotted owl was still a local issue, Forest Service leaders might have achieved a resolution without compromising their mission.[535] Instead, the agency lost both control over the issue and credibility with the other parties.

In 1988, Oregon and Washington listed the spotted owl as threatened and endangered under state laws, which placed pressure on the U.S. Fish and Wildlife Service to list the owl for protection under the federal Endangered Species Act. In 1989, an Interagency Scientific Committee (ISC) was established to develop a "scientifically credible" conservation strategy. Forest Service researcher Jack Ward Thomas chaired the committee. A well-respected wildlife biologist, Thomas had led the ecosystem research used to develop guidelines for managing wildlife and timber on four national forests totaling four million acres in the Pacific Northwest in the late 1970s. This was the first widescale application of an ecosystem approach, one that was widely studied and used to guide management throughout the National Forest System.[536] Eric Forsman and four other researchers joined Thomas on the committee. Before the spotted owl controversy subsided three years later, Thomas would participate on four blue-ribbon panels, all assembled to address the same fundamental question— what was the greatest good for the greatest number in the long run?

While the ISC studied the issue, mainstream environmental groups conducted lobbying and letter-writing campaigns, but radical environmental groups like Earth First! and Greenpeace became increasing confrontational with both timber companies and governmental agencies, driving metal spikes into trees to make it dangerous to cut into the wood and chaining themselves to trees marked for cutting. Their tactics, which destroyed private property and caused injury, earned them the enmity of the Forest Service and the timber industry.[537] Timber industry workers blamed both environmental activists and the government for their economic woes, and timber industry leaders pressured the Forest Service to get the cut out. Chief Robertson and Jack Ward

After studying the northern spotted owl, the Interagency Scientific Committee concluded that conservation in the Pacific Northwest was "more complex than spotted owls and the timber supply— it always has been." (Tom Iraci – USDA Forest Service)

As Robertson's successor as chief, Jack Ward Thomas took over the task of implementing ecosystem management. Thomas explained the difficulty of trying to implement the approach:

"Ecosystem management is a concept, and concepts are always fuzzy. In concept, we are going to think at broad scale, across boundaries, for a longer time frame, [and] include an expanded number of variables, ranging from social and economic to biological and ecological. When the concept is applied to a specific place" and specific time frames with these specific variables to be considered, then the concept is "in context and no longer nebulous. All concepts must be placed into context before they become meaningful—and that is not just in the case of ecosystem management.... Multiple use is nebulous until you define it."

Source: Steen, Interview with Jack Ward Thomas, *69*.

Thomas received death threats and were assigned bodyguards, usually from the Forest Service's Law Enforcement division.

Thomas knew that the spotted owl investigation would change his professional life, but it disrupted his private life, too. In his first job, working with the Texas Parks and Wildlife Department in the 1960s, Thomas had received death threats but none that he took too seriously. The spotted owl situation was another matter. At one point, the Federal Bureau of Investigation advised Thomas not to return to his home and placed him and his family in protective custody. One morning he was about to get into his truck when he noticed that the hood was ajar. When he saw what he thought was dynamite, he called the authorities, who found three road flares on the engine block. They were not wired, Thomas recalled, "but the point was well made."

After the committee issued its report in April 1990, Thomas received a threatening call in the middle of the night. He interrupted the caller's litany of threats and told him to "get a job" and hung up the phone. The phone rang again. Thomas, realizing the caller was a little drunk, informed him that he did not take death threats at home. "We receive death threats at the office between eight and five" during the workweek. He then gave his office phone number, even taking the time to repeat it "very slowly and distinctly," before hanging up.[538]

Environmental critics feared that the committee would hand the government scientific justification for what it had been doing, and that in a battle of "jobs versus owls," jobs would win. Instead, the ISC scientists stated that conservation in the Pacific Northwest was "more complex than spotted owls and the timber supply—it always has been." Later investigations showed that mechanization and other technological changes in the timber industry, along with the export of whole unfinished logs to foreign markets, had led to the loss of some timber industry jobs long before the spotted owl became an issue.[539]

The ISC report cited many factors that "must be considered when evaluating the conservation strategy," including water quality, fisheries, soils, wildlife, and outdoor recreation.[540] The report proposed a two-part conservation strategy: first, steps to ensure the owl's long-term survival, and then "research and monitoring to test the adequacy of the strategy and to seek ways to produce and sustain suitable owl habitat in managed forests."[541]

The Bush administration found politically untenable the call for protecting landscapes and eliminating timber harvests until protective silvicultural methods could be developed. The administration appointed a task force, again led by Jack Ward Thomas, to develop alternatives that would allow cutting to continue. The Forest Service agreed to abide by the findings of the Thomas Report, but the Bureau of Land Management invoked the "God Squad" provision of the Endangered Species Act, which allows letting a species go extinct if the social and economic costs are deemed too high. BLM hoped that pushing forward its plans for forty-four of fifty-two timber sales in areas that the Fish and Wildlife Service had called critical habitat would pressure Congress to amend the act, and that would allow timber harvesting to resume at the higher rates.

With their livelihoods threatened if the Forest Service shut down logging operations in the Pacific Northwest, loggers staged protests in support of continued clearcutting on the Region 6 national forests in the 1980s. (Jim Hughes – USDA Forest Service)

Many of the sales went through (though no timber was ever cut), but the issue became a major point of contention in the 1992 presidential election. One month before the election, the federal government listed the marbled murrelet, a robin-sized seabird that nests in Pacific coastal forests, as endangered, and stocks of Pacific salmon and steelhead were dangerously low. More than 480 species were considered to be of concern in the region's late-successional forests. By election time, roughly 160 were candidates for federal listing as threatened or endangered.[542]

William J. Clinton, who had promised in the presidential campaign to hold a summit to resolve the issue if elected, made good on his promise in April 1993. The summit brought together President Clinton and Vice-President Al Gore, several cabinet members, and environmentalists, timber workers, and community representatives. Leaders of the Forest Service and Bureau of Land Management, the two agencies responsible for managing the public lands in question, were not invited to speak, a clear indication that they were out of favor with the new administration. At the end of the summit, President Clinton directed Thomas and the newly established Forest Ecosystem Management Assessment Team (FEMAT) to create a legally, socially, and ecologically viable plan for the region's forests. Signed by the president in spring 1994, FEMAT's Northwest Forest Plan environmental impact statement was challenged in court by both environmental activists and the timber industry. When it received approval from the courts in December 1994, five years of lawsuits and "forest gridlock" ended.

The plan represented the first time that the Forest Service and the Bureau of Land Management, along with the National Marine Fisheries Service, the Fish and Wildlife Service,

The "timber summit" held by President Bill Clinton soon after he took office in 1993 broke the "forest gridlock" in the Pacific Northwest, but it did not resolve the Forest Service's timber management problems elsewhere. Controversy continues to surround many of the agency's land management policies and decisions. (White House)

and the Environmental Protection Agency, agreed to a common management approach for an entire ecological region. The Northwest Forest Plan established a series of habitat conservation areas and reserves (called allocations) for species dependent on old-growth and established similar protection for aquatic species in western Washington and Oregon and northwestern California, an area of approximately 24.5 million acres. The accompanying standards and guidelines formed a comprehensive management strategy that included protection measures, restoration activities, and commercial timber harvest.[543]

Logging continued after 1993 in the national forests of western Oregon and Washington but at about one-fifth the level in the 1980s. By then, much of the forest cover lacked the complex composition favored by the northern spotted owl. As the FEMAT report noted, it took nearly a century and a half to reduce the forest to this simplistic composition, and it was going to take much longer than a mere decade or two to bring it back.[544]

ICBEMP

The strategy developed under the Northwest Forest Plan served as a prototype for other regions around the country. The Interior Columbia Basin Ecosystem Management Project (ICBEMP), formed in 1993, took nearly ten years to complete. The project assessment area included more than 140 million acres in eastern Washington and Oregon, Idaho, and western Montana. The management strategy, however, applied only to the approximately 64 million acres of public lands administered by the Bureau of Land Management and Forest Service within this area.

ICBEMP well illustrated the difficulties of trying to manage public lands for divergent constituencies and without deadlines. In summer 1995, following the November 1994 congressional elections that produced a Republican majority, western Republicans attempted to force BLM and the Forest Service to generate seventy-four management plans—one for each BLM district and national forest in the Columbia basin—and end implementation of two interim plans designed to protect fish habitats. They argued that managing such a large ecosystem plan would infringe on private property rights, even though individual plans would cost more time and money to develop and might jeopardize ecosystem health in the meantime. When that effort failed, Republican leaders proposed to cut funding from the $5.7 million needed to complete ICBEMP to only $600,000. The federal budget impasse between the president and Congress in late 1995, however, put the Republicans on the defensive. After congressional representatives from the affected districts in Oregon and Idaho held hearings and entered statements into the record for their constituents back home, the proposal died and ICBEMP proceeded.[545]

The two draft environmental impact plans that were released in June 1997 prompted more than 83,000 public comments during the 335-day comment period. Addressing those comments, as well as new scientific information, agency review, and direction from the secretaries of Agriculture and Interior, a supplemental draft was released in March 2000; it drew some 525 comments over a 90-day comment period. The project team's final environmental impact statement and proposed decision in December 2000 initiated a protest process that lasted for a month; 74 protest letters were received. Nearly ten years after the project began, representatives of the Forest Service, Bureau of Land Management, Fish and Wildlife Service, National Marine Fisheries Service, and Environmental Protection Agency signed a memorandum of understanding, completing the planning phase. In contrast, FEMAT, which started with a defined deadline, completed the Northwest Forest Plan project in two years.

In 1998, partly inspired by the promise of the Northwest Forest Plan and ICBEMP, the Forest Service launched other large-scale ecosystem assessments for California and Nevada and for the South. Unlike ICBEMP and the Northwest Forest Plan, the Sierra Nevada and Southern Appalachian plans generated only assessments, not alternative management plans, and consequently generated little controversy.[546]

A Reluctant Chief in the Political Whirlwind

Chief Robertson is rightfully credited with refocusing the agency's management objectives through his decision to support and implement the New Perspectives ecosystem management program.[547] But President Bill Clinton's team viewed him as a Reagan man who did not agree with what the new Democratic administration hoped to accomplish. Robertson endured perhaps the most criticism of any chief, much of it centered on how the Forest Service handled the northern spotted owl and old-growth forest controversy in the Pacific Northwest, and he knew that with the Clinton-Gore victory, his days were numbered. Robertson and Vice-President Gore had had disagreements while Gore was a U.S. senator. It came as little surprise to Robertson that he was not asked to speak at the Northwest timber summit in April 1993.[548]

With the Clinton administration eager to demonstrate its commitment to promoting environmental issues, Robertson was ousted later that year. Jack Ward Thomas, who had achieved national prominence during his work on the spotted owl blue-ribbon panels, became the leading candidate to replace him. Anticipating the demands the job would place on him and his family, Thomas did not want the appointment. He preferred to return to his wildlife research projects, and he resented the poor treatment that Robertson had received as chief.[549]

Thomas also hesitated because he was not a member of the federal government's Senior Executive Service. His appointment would have to be a political appointment instead of a conventional promotion and would rankle those above him. Of greater concern, accepting a political appointment would set a precedent and open the Forest Service to more political appointees. The administration hinted that if he did not accept, it would go outside the agency for a new leader—which Thomas thought would further demoralize an agency beset by controversy—but if he did, Thomas's political bosses promised, they would convert his appointment to a Senior Executive Service position. When he reluctantly accepted, believing it for the good of the agency,[550] Thomas became the first wildlife biologist and the first nonforester and nonengineer to serve as chief. His position was never converted from a political appointment—a move that the promise makers lacked legal authority to effect.

Thomas took over an agency in tumult and transition. The Forest Service faced public opposition to practically anything it proposed. The handling of roadless areas and wilderness preservation, logging, mining, and livestock grazing—issues that had characterized public land management for the last forty years—had to be addressed if both environmental and day-to-day working conditions were to improve. In the Pacific Northwest, for example, Thomas and BLM Acting Director Michael P. Dombeck called for strategies that would produce a "quantum leap forward" in protecting and restoring fish habitats. The two men worked closely together on similar issues.[551]

Thomas's plainspoken, direct manner won him support in some quarters of the agency as well as a few enemies at both ends of Pennsylvania Avenue. He had to contend with a Forest Service sometimes at odds with itself over the direction it was taking, a timber industry outraged over the dramatic drop in timber production in the Pacific Northwest, a skeptical public, a White House that did not seem to fully appreciate the agency's complex situation, and a Congress that was interested in rolling back regulations and keeping the agency on a tight budget. The Republican-controlled Congress in the 1990s, especially members from timber-dependent western states, considered Thomas the "spotted owl guy" or the "Clinton guy," which made his dealings with Congress difficult. Some were unhappy that Thomas continued the painstaking

implementation of ecosystem management on the national forests and grasslands. In Thomas's estimation, they took their anger out on him and the Forest Service for the decline in western timber harvests.[552]

One of the enduring myths about Forest Service chiefs is that they are insulated from politics. Gifford Pinchot's dismissal provides one proof to the contrary; the salvage rider during Thomas's term is another. After the 1994 fire season, when a record number of firefighters died and 1.4 million acres of forest burned, Congress decided to spend less on fire suppression and more on prevention. President Clinton signed the 1995 Rescission Act, an emergency appropriations bill that provided funding for several agencies, including the Forest Service. The Forest Service received funding for the Emergency Salvage Timber Sale Program, better known as the salvage rider.

The salvage rider attempted to give the Forest Service more leeway in promoting "forest health." Environmental advocates might judge forest health by species diversity or the viability of populations of certain species; timber companies see indicators of forest health in the amount of timber available for harvest. In this case, forest health was assessed to be at risk because the dead and dying trees in western forests presented fire and disease hazards.[553] The salvage rider exempted certain timber sales in burned and diseased forests from environmental constraints, such as riparian buffer zones and habitat protection, and allowed the removal of green trees as well to reduce fuel loads. Environmental activists were outraged. Thomas did not support the bill, believing that the agency stood to lose no matter how it interpreted it. Once it passed,

A member of the Asheville (North Carolina) Hotshots ignites a prescribed fire at the Department of Energy's Savannah River site in South Carolina in 2004. Ecosystem research overturned long-held assumptions regarding the role of fire in maintaining healthy ecosystems. (Kim Ernstrom – USDA Forest Service)

though, he proceeded with the sales but ordered forest managers to follow the environmental laws anyway, including consulting with all the appropriate agencies, such as U.S. Fish and Wildlife Service. His decision to carry out the sales in this manner pleased no one.[554]

The program, involving the removal of 4.5 billion board feet of salvage timber, was under way when the 1996 presidential campaign began. In response to complaints from the environmental community, the White House ordered Thomas to slow down, which outraged timber industry supporters. As the election drew closer and the outcome in the Pacific Northwest grew increasingly doubtful, the White House switched its position to court timber votes and criticized Thomas for being too slow, which led environmental activists to attack the agency. As a chief committed to obeying the law and telling the truth, Thomas was angered and deeply hurt to see the Forest Service used as a political pawn.[555] The target was met and the controversy subsided, but forest health issues would resurface in 2003 with President George W. Bush's Healthy Forests Initiative.

Relations between Forest Service leaders and the administration deteriorated further. Thomas was asked by the White House to relieve five of his top staff, which he refused to do, and on another occasion, he was told that the White House did not consider him "one of them." In December 1996, a month after Clinton's reelection and nearly three years to the day after his appointment, he resigned, having served the shortest tenure of any Forest Service chief.[556] He followed tradition and provided a list of people, including Michael P. Dombeck of BLM, for the secretary of Agriculture to consider as his replacement. Dombeck was appointed in January 1997.

Dombeck's Natural Resource Agenda

Mike Dombeck had joined the Forest Service in 1978 as a fisheries biologist and spent almost twelve years with the agency. In 1989, at Chief Robertson's request, Dombeck accepted appointment as science adviser and special assistant to the director of the Bureau of Land Management. In February 1994, he was named acting director of BLM, a position that allowed him an opportunity to foster closer cooperation between his agency and the Forest Service and which, later, as Forest Service chief, gave him insight into how to bring the two agencies closer together.

Before taking office, Dombeck held meetings and conference calls with the agency's Washington office leaders, the regional foresters, station directors, and others to discuss the sour relations with the White House and Congress, and between those in the field and the Washington office. He found a lack of trust all the way around. The top staff members that Thomas had been asked to dismiss expressed little desire to work with the White House to resolve differences. When Dombeck tried to transfer some staff members and bring in new personnel in hopes of improving relations and the Forest Service's standing with the White House, the move was dimly viewed by many in the agency. Dombeck had gotten off to a rocky start.[557]

A year after becoming the first Forest Service chief in sixty years appointed from outside, Dombeck introduced his Natural Resource Agenda to steer discussion back to conservation values. To recapture the mantle of conservation leadership, he refocused the Forest Service's work on conservation problems. He did so with the encouragement and support of the secretary of Agriculture, and with the input of his own national leadership team. Dombeck declared in issuing his agenda, "We have two very basic choices. We can sit back on our heels and react to the newest litigation, the latest court order, or the most recent legislative proposal. This would ensure that we continue to be buffeted by social, political, and budgetary changes." Dombeck proposed going in a different direction. "[W]e can lead by example. We can lead by using the best available scientific information based on principles of ecosystem management to advance a new agenda. An agenda with a most basic and essential focus—caring for the land and serving people."[558]

That slogan—caring for the land and serving the people—had been articulated in the mid-1980s, while Max Peterson was chief, but it had been lost in "analysis paralysis" and all the other problems that had beset the agency ever since. Dombeck hoped to generate positive press about the agency with his resource agenda. According to Dombeck, it worked. He estimated that coverage of the agency went from being eighty percent negative to eighty percent positive.[559]

The Natural Resource Agenda had four areas of emphasis: watershed health and restoration, a long-term forest roads policy, sustainable forest management, and recreation. The most controversial part of the agenda was the roads policy, or more precisely, the roadless areas policy. The idea stretched back to the 1920s, when Arthur Carhart had proposed setting aside wilderness

Researchers had spent decades conducting experiments at ground level until facilities like the Wind River Canopy Crane Research Facility was established in 1994 to study biological and ecological processes at all levels of an old-growth forest, from the atmosphere to below ground. The crane allows scientists to comfortably work at treetop level. Located on the Wind River Experimental Forest in the Gifford Pinchot National Forest (Washington), it is closely allied with the H. J. Andrews Experimental Forest. (Tom Iraci – USDA Forest Service)

areas. Until 1964, when Congress passed the Wilderness Act, the agency had managed nine million acres as wilderness areas. The Wilderness Act ordered the agency to review roadless areas larger than five thousand acres to determine their appropriateness for inclusion in the National Wilderness Preservation System. The Roadless Area Review and Evaluation (RARE) findings of 1972 designated approximately fifty-six million acres as wilderness, all in the West. The Forest Service was ordered to complete another such review in 1979. RARE II, which included eastern forests after the Forest Service loosened its definition of wilderness, recommended in 1979 that Congress add fifteen million acres to the National Wilderness Preservation System.[560]

A road brings the competing needs of forest users into conflict. Loggers and many recreational users benefit from roads, but those who seek a more primitive experience in the woods do not want them. Many forest supervisors, to avoid negative publicity and backlash from environmental groups and some politicians, avoided implementing proposals to build roads. Proposed timber sales in roadless areas, which provoked controversy because they would require the construction of roads into those areas, prompted Chief Thomas to order his forest supervisors in 1995 to either move forward with the harvest plans or remove those areas from consideration.[561] Thomas hoped his orders would bring the controversy to a head. He was right. Congress nearly cut the Forest Service's road budget by eighty percent as a way to protect the roadless areas and preserve wilderness. Had it done so, it would have crippled the agency's ability to manage the land.

Roadless areas continued to be an issue for the Forest Service. In January 1998, Chief Dombeck announced plans to suspend new road construction in most inventoried roadless areas. Thirteen months later, he proposed suspension of new road construction in roadless areas for eighteen months. Exceptions included Alaska's Tongass National Forest and Pacific Northwest forests, which had recently revised forest management plans. Some environmentalists wanted those areas protected by the new policy as well.

Dombeck's objective was to give the Forest Service time to develop a new long-term forest road policy in light of an $8.4 billion road maintenance backlog for the 375,000 miles of roads in the National Forest System. (The interstate highway system has only 43,500 miles of roads.) In 1989, forty-seven percent of forest roads were being maintained to standard; by 1997, only thirty-eight percent. Falling timber sales in the 1990s had reduced funds for road reconstruction, and impassable roads meant that Americans could not safely navigate their public lands. Not building new roads when existing ones needed repair seemed logical.[562]

The announcement took forest supervisors by surprise, and many were unprepared to respond to the press. Dombeck believed that making the decision *in camera* was the best way to avoid "a long protracted dialogue about whether or not we should do it" and prevent having the decision removed from the Forest Service. In hindsight, Dombeck admitted that perhaps he should have given his forest supervisors and district rangers advance notice, but he maintained that making the policy effective was worth their criticism.[563]

Timber interests were surprised, too, and angered. Dombeck's decision to stop road construction had halted opening new areas for timber harvesting. Four outraged congressional leaders from the timber states of Idaho and Alaska—known within the Forest Service as the Four Horsemen of the Apocalypse—wrote a letter threatening to slash the agency's budget to a "custodial" level because the Forest Service was not harvesting timber at levels they wished to see. As Dombeck wrote later, "With Congress unable to pass new legislation or otherwise resolve the controversy, with litigation determining more and more how national forests would be managed, Forest Service leaders believed it was up to them to resolve the issue." Congressional representatives's threats "to take their football and go home" proved empty. Instead, the road suspension helped put the Forest Service back in the land management game.[564]

To help shape its roadless areas policy, the agency held 187 meetings attended by more than sixteen thousand people around the country and, over the course of the two-year process, received more than a million comments on the issue. Poll after poll showed that the majority of the people favored the suspension of new construction. For the first time in years, the agency enjoyed active support from environmental advocacy groups. As it became evident that the Forest Service also favored retaining the right to conduct timber harvesting as a management tool in the future, support in some environmental quarters melted away. Yet the Forest Service's

In February 2004, Chief Dale Bosworth (center, facing camera) received a field briefing from local, county, and agency officials coping with a beetle infestation that destroyed timber on and around the San Bernadino National Forest. As land management issues have grown more complex, the Forest Service has worked more closely with local, county, state, and other federal agencies to address them. (Tom Iraci – USDA Forest Service)

willingness to raise an issue certain to ignite passions and then discuss it was a step toward reclaiming the agency's traditional position as a conservation leader for the federal government.

The Forest Service's new roadless areas policy put Chief Dombeck at odds with the presidential administration of George W. Bush as it took office in early 2001. Dombeck, who did not want to compromise conservation values that he knew conflicted with the incoming president's, submitted a list of possible successors, and Dale N. Bosworth was chosen to succeed him.[565] Shortly after Bosworth took over, the administration rescinded Dombeck's roadless area decision as a national policy and Bosworth returned the decision-making powers over road construction to his regional foresters.[566]

Dale Bosworth and The Four Threats

A second-generation forester and Forest Service employee who had been in and around the Forest Service all his life, Dale Bosworth began his Forest Service career in northern Idaho in 1966. He served as regional forester for Region 4 (Intermountain Region) from 1994 to 1997 before becoming regional forester for Region 1 (Northern Region) in August 1997. Though he was nearing retirement age and had minimal experience with Washington politics, he accepted the appointment out of a sense of duty to the agency.[567]

A lifetime of experience and a career largely spent in the field shaped Bosworth's thinking about what he hoped to accomplish as chief. The changes in the Idaho landscape that occurred between his first job in Idaho as a forester and his return there thirty years later as regional forester led him to reconsider the impact of human actions on national forests in general.[568] As chief, he wanted to focus on four threats to ecosystem health—fire and fuels, invasive species, loss of open space, and unmanaged outdoor recreation. How to manage the forests and grasslands had largely been settled in the early 1990s in favor of ecosystem management and ecosystem restoration. The days of high timber harvest levels on the national forests were gone. Bosworth now wanted the agency to decide *what* to manage. The debate should be about not what is taken from the forest but what is left behind; it was no longer about outputs but about outcomes.[569] By reframing the debate, Bosworth hoped to get the Forest Service's focus and resources out of the courtroom and into the field. It was a different way of asking, What is the greatest good?

When he proposed the "four threats" agenda to his staff, some staffers wanted to use a more positive word, such as "opportunity." Bosworth believed that these four problems were indeed threats and should be labeled as such if the agenda were to become a subject for public dialogue.[570] In a series of speeches in 2003 and 2004, he declared that invasive species (both plants and aquatic and terrestrial animals) posed a threat because "where the ecological controls they evolved with are missing," they "crowd out or kill off native species, destroying habitat for native wildlife. Where cheatgrass takes over, for example, the range loses forage value for deer and elk." Kudzu, an invasive plant introduced from Japan in 1876 and used for soil erosion control

and fast-growing forage beginning in the 1930s, now covers more than seven million acres. Kudzu and other vines such as Japanese honeysuckle and English ivy can retard the growth of seedlings and reduce timber yields and have fundamentally altered the character and structure of some forests. Estimates showed that invasives cost taxpayers about $137 billion per year in economic damages and associated costs.[571]

The loss of open space to development—estimated in the late 1990s at a rate of about four thousand acres every day and increasing—affects wildlife habitat and the quality of human life. Development contributes to habitat fragmentation, ownership fragmentation (the conversion of large acreages into smaller parcels), and use fragmentation (the transformation of large, single-use tracts used for forestry, farming, and ranching into small, multiple-use tracts). Those different forms of fragmentation can contribute to overall ecosystem degradation.

"Where private open space is lost, recreational pressures on public lands tend to grow," Bosworth noted. Those pressures are keenly felt on the national forests and grasslands, contributing to another threat—unmanaged outdoor recreation. The number of off-highway vehicle users grew from about five million in 1972 to almost thirty-six million in 2000. "Each year," Bosworth noted, "the national forests and grasslands get hundreds of miles of unauthorized roads and trails due to repeated cross-country use. We're seeing more erosion, water degradation, and habitat destruction. We're seeing more conflicts between users."

The wildland-urban interface is increasingly an area of concern for the Forest Service. Houses built in and around national forests are at risk during fire season, but they also have an impact on wildlife because forest fragmentation, construction, and occupation can disturb habitat. (Tom Iraci – USDA Forest Service)

Of the four threats, Bosworth gave greatest emphasis to fire because of its direct impact on a wide range of landscapes, from primitive wilderness to urban forests, and on both public and private lands. It also affects the many communities in the wildland-urban interface. The destructive fire seasons of 2000 (more than 8 million acres of forest and 861 homes and structures burned) and 2002 (more than 6.9 million acres and 815 homes and structures) gave impetus to his agenda. After President Clinton toured fire damage in Montana in 2000, bipartisan cooperation on Capitol Hill led to an increase in the fire budget to fund an interagency effort, labeled the National Fire Plan 2000. The plan was designed to deal with fuel buildups on federal lands by rehabilitating and restoring lands and communities affected by fire, and using techniques such as prescribed fire to reduce hazardous fuels.[572]

The National Fire Plan was a step forward, but the fuel buildup problem remained. Bosworth explained the situation: "Many of our fire-dependent ecosystems have become overgrown and unhealthy. The answer is to reduce fuels before the big fires break out. Where fire-dependent forests are overgrown, we've got to do some thinning, [and] then get fire back into the ecosystem when it's safe." Bosworth offered the example of using more prescribed fire in shrubby systems such as chaparral in southern California "to take some of the heat out of those systems."[573]

President Bush's administration used Bosworth's discussion of fuel reduction in crafting the Healthy Forests Restoration Act of 2003, which Congress passed after the southern California fires in the autumn of 2003 had drawn national media attention to the issue. Healthy Forests was designed to give the Forest Service flexibility in working to reduce fire and fuel loads by streamlining regulations to remove hazardous fuels on public lands. In January 2005, the Bush administration issued new regulations developed by the Forest Service for the National Forest System's land management planning framework. The new rules restructured the process by shifting from prescriptive to strategic planning, to give forest planners and supervisors greater flexibility in responding to environmental, social, and scientific changes. Forest Service officials estimated the new regulations would cut planning costs by thirty percent and would allow forest managers "to finish what amount to zoning requirements for forest users in two to three years, instead of the nine or ten years they sometimes take now."[574]

Bosworth, who had been involved in the development of regulations at the end of the Clinton administration, believed that the Forest Service needed to update its regulations to reflect forest conditions. Previous regulations had been designed to manage the lands during the heyday of high timber harvests, when national forest timber harvest levels rose by 92 percent between 1952 and 1986 and peaked at about fourteen billion board feet. Between 1986 and 2001, they declined by 84 percent as timber harvesting shifted from public lands in the Pacific Northwest to non-industrial private timberlands in the South following the spotted owl decision. By the late 1990s, timber harvests averaged only about two billion board feet. In 2001, national forests accounted for only two percent of the nation's timber harvest, the same level they had been between 1905 and World War II.[575]

By the time Bosworth took over in 2001, the Forest Service was fully committed to ecological restoration and outdoor recreation but remained mired in litigation.[576] Bosworth's Four Threats agenda and his involvement in drafting new regulations for cutting aimed to keep the Forest Service in a leadership position in federal conservation efforts by enabling the agency's experts to practice conservation in the field.

Like most changes affecting how the Forest Service operates on public lands, Healthy Forests and the January 2005 planning regulations have met with a mixture of criticism and praise. Environmental groups expressed outrage over the new regulations, arguing they would limit or remove environmental protections and public participation from the forest planning process. Those in the timber industry still logging on national forest lands and many people living adjacent to fire-susceptible national forests supported the measure. The same is true of the rescinded roadless area policy—Bush's new policy angered most environmental advocates but pleased segments of the timber industry and some recreational users. Once again, in trying to address the question of what is the greatest good, the Forest Service took into account the needs of competing constituents while doing what its own leaders believed was best for the land in the long run.

Chapter Ten

REFLECTIONS ON
THE GREATEST GOOD

Forest Ranger Griffin and Forest Guard Cameron on fire patrol duty, Cabinet (now Lolo) National Forest, 1909. (USDA Forest Service – Forest History Society)

SHORTLY AFTER TAKING OVER AS REGIONAL FORESTER FOR THE PACIFIC NORTHWEST IN 1992, JOHN LOWE NOTED, "THERE WILL BE LOTS OF DIFFERENT WAYS OF MAINTAINING SUSTAINABLE ECOSYSTEMS, AND THE METHODS FOR DOING THAT WILL GENERATE CONTROVERSY. ECOSYSTEM MANAGEMENT WON'T TAKE THE CONTROVERSY OUT OF THE FOREST."[577] IF ONE SUBSTITUTES "FIRE MANAGEMENT" OR "RANGE MANAGEMENT" OR "WATERSHED PROTECTION" FOR "ECOSYSTEM MANAGEMENT," LOWE'S ASSERTION CAN BE APPLIED AT ANY POINT IN THE HISTORY OF THE FOREST SERVICE. AFTER ALL, THE AGENCY WAS BORN IN CONTROVERSY.

Over the decades, there has been no shortage of controversy regarding public lands. The writings of George Perkins Marsh and others predicting the demise of the nation if its land was not handled properly were antithetical to the policy of dispensing the public lands as quickly as possible. Reaction to the Forest Reserve Act of 1891 and President Cleveland's creation of the Washington's Birthday forest reserves led to the Organic Act in 1897, a controversial measure in its own right that laid down the first guidelines of federal forestry management. Control of the forest reserves remained with the Department of the Interior, however, while the government's foresters resided in the Department of Agriculture's Division of Forestry.

The Division of Forestry grew rapidly under the leadership of the dynamic Gifford Pinchot, who seized every opportunity to demonstrate both the need for scientific forest management and the competence of his men to carry it out. He established the Yale Forest School to train foresters, developed the division to employ the trained foresters and give them experience, and created the Society of American Foresters to foster the exchange of professional ideas and research findings. Fulfilling Pinchot's vision, forestry took the lead in the growing conservation movement, and his agency quickly became the paragon of government land management. Within seven years, Pinchot, with the support of Overton Price, Albert Potter, and others, had prepared the agency to manage the forest reserves when they were transferred to the Department of Agriculture.

After much debate and discussion, the executive and legislative branches came together to establish the Forest Service in 1905. Conflict surrounded its establishment, and resolving conflict became one of its purposes. Pinchot addressed the issue when he crafted the Forest Service's mission statement: "Where conflicting interests must be reconciled, the question will always be decided from the standpoint of the greatest good of the greatest number in the long run." Pinchot defined the agency's mission broadly so that its future leaders and employees would have the flexibility to respond as environmental and political conditions changed.

Although a major reason for the agency's establishment was to stave off a timber famine, little logging occurred on the national forests until the 1920s. In the meantime, Pinchot and his two immediate successors, Henry Graves and William Greeley, focused on establishing policies for other land uses, such as grazing, and the authority to carry out those policies. The courts upheld the agency's right to manage the land by its imposing regulations and fees on users. Early debates over grazing and fire policy, along with the rivalry with the National Park Service over recreation and land-use policies, helped shape and establish most Forest Service policies.

Though policy debates continued into the 1910s and 1920s, they were largely settled by the time of the Great Depression and the New Deal. New national economic and ecological crises challenged the Forest Service. Making effective use of the Civilian Conservation Corps and other government programs, the Forest Service worked to save imperiled natural resources. Given the men and means to carry out their conservation plans, Forest Service leaders remedied old problems like soil erosion in the Great Plains and fire in the forests while rehabilitating the recently acquired grasslands and eastern national forests. World War II upended Forest Service priorities, including its push for regulating private lands, and that same energy and zeal now went toward winning the war by meeting escalating demands for lumber. Timber management, which had been the Forest Service's focus since its inception, soon defined all its efforts in the Cold War. Economic needs trumped ecological concerns and recreation demands in the national forests, but few understood the ramifications of the decision to get the cut out.

The concept of the greatest good began to undergo scrutiny, and new voices began to ask for a seat at the table where it was defined. Legislation such as the Multiple Use–Sustained Yield Act and the Wilderness Act tried to restore balance between timber and other uses, but by then even-age management had become the standard practice on the national forests, and the agency had earned a sterling reputation. Confidence in its abilities to handle land management problems blinded both agency leaders and their political bosses to changing public attitudes and new scientific findings. The Forest Service's failure to recognize or give credence to such changes led to the Bitterroot and Monongahela decisions, which began the agency's legal woes and the analysis paralysis that hobbled the agency into the 1990s.

Debate and discussion, absent from the Forest Service at the heights of its power and popularity at midcentury, reentered the halls of the agency in the 1970s. No less than their "old guard" supervisors, new employees who came from different professional and cultural backgrounds believed in the Forest Service and its position as a leader in conservation and wanted to reestablish the agency's credibility with the public. But conflict and disagreement over approaches to land management produced more than rancor within Forest Service ranks in the 1970s and 1980s.

Many of the laws that opened doors to women, minorities, and nonforesters in the agency also unexpectedly opened the doors to land management through litigation. As Congress and the White House pressed the agency to raise timber production, the public split over the preferred uses for national forests. Private industry relied on the national forests for commodities such as timber, grazing, and paid recreational activities like rafting or fishing, and demanded increased production levels. Environmental activists and other concerned citizens wanted to curtail or end extractive activities—chiefly timber harvesting—and recreational activities like snowmobiling and mountain biking in many of these areas. Some wanted minimal human intervention instead. Meanwhile, scientific research had revealed that a mix of human and natural solutions could restore ecological health and still provide commodities.

The Forest Service, however, could not implement new management approaches with the ease it once had. The prescriptive laws, the size of the agency, and savvy, litigious advocacy groups constrained the agency's latitude in dealing with land management issues in the 1980s and 1990s. The northern spotted owl, whose imperilment effectively ended timber harvesting on the national forests in the Pacific Northwest, gave environmental advocates a victory, and the Forest Service paid the price for having ignored those opposed to its practices twenty to thirty years earlier. The old ways no longer worked—forestry as it had been practiced for the previous half-century had hit the wall.

As the Forest Service struggled to regain its leadership in the conservation arena, it turned to ecosystem management. An on-the-ground embodiment of Aldo Leopold's land ethic, ecosystem management held out the promise of placing both commodities and ecosystem health on a sustainable basis. Forest Service leaders starting with Dale Robertson asked their employees to return their attention to the forest. Still impeded by competing interests and contradictory laws, the Forest Service asked again, "What is the greatest good?" and reinvigorated the debate and discussion about the public purpose of the national forests.

Those two elements—debate and discussion—are essential to determining how to manage the public lands one hundred or more years into the future. The national forests and grasslands are, after all, the public's lands. In the absence of debate and discussion, the Forest Service went largely unchallenged in the 1940s and 1950s and, in the process, unknowingly put everything it had stood for at risk. In its most recent efforts to restore the land, agency leaders have been engaging in discussion and debate with its employees and with the public it serves. The discussion will need to be far ranging as the nation and the Forest Service seek to restore and maintain the ecological health of the public lands. Like the land itself, the challenges are complex—and perhaps more complex than we can now comprehend.

That discussion will address the meaning of "the greatest good for the greatest number for the long run." Gifford Pinchot did not intend that statement as policy but as a framework for discussion and a guide for considering policy. To have meaningful and productive dialogue, we must continue to ask ourselves, What is the greatest good?

Interior of an office tent, Stanislaus National Forest (date unknown). (USDA Forest Service)

Appendix A

Chronological Summary of Events Important to Forest Service History

1862 Department of Agriculture is established (12 Stat. 387).

1864 George Perkins Marsh publishes *Man and Nature.*

1872 Mining Law (30 U.S.C. 22, 28) is passed, allowing prospecting and mining of hard-rock minerals on national forests and grasslands.

1873 American Association for Advancement of Science resolves in favor of creating federal forestry commission; Franklin B. Hough heads committee to implement resolution.

1875 American Forestry Association (now American Forests) is organized in Chicago.

1876 Congress appropriates $2,000 to employ federal forestry agent (19 Stat. 143, 167); Franklin B. Hough is appointed.

1881 Division of Forestry is established in Department of Agriculture.

1882 First American Forestry Congress is held in Cincinnati.

1883 Nathaniel H. Egleston succeeds Hough as chief of Division of Forestry.

1886 Bernhard E. Fernow succeeds Nathaniel Egleston as chief of Division of Forestry. Division receives statutory permanence within Department of Agriculture (24 Stat. 100, 103).

1889 Department of Agriculture receives cabinet status.

1891 President is authorized to set aside forest reserves from public domain (26 Stat. 1095).

1892 Sierra Club is formed.

1896 National Academy of Sciences appoints special committee to investigate forest reserves; Charles S. Sargent is chairman.

1897 Amendment to Sundry Civil Appropriations Act (30 Stat. 11, 34, Organic Act) specifies purposes for which forest reserves can be established and provides for their administration and protection.

1898 Gifford Pinchot succeeds Bernhard Fernow as chief of Division of Forestry.

New York State College of Forestry and Biltmore Forest School open.

1900 Yale Forest School opens.

Society of American Foresters is formed.

Lacey Act (16 U.S.C. 701) authorizes secretary of Interior to adopt measures to restore game birds and other birds where they have become scarce or extinct and to regulate new species introductions.

1901 Division of Forestry created in General Land Office with Filibert Roth at its head.

Division of Forestry in Agriculture becomes Bureau of Forestry (31 Stat. 922, 929).

1902 National Lumber Manufacturers Association is formed.

1905 Second American Forest Congress held in Washington to pressure Congress to support transfer of forest reserves.

Administration of forest reserves is transferred from Interior to Agriculture, effective February 1 (33 Stat. 628).

1905 Law Enforcement Authority (under Agricultural Appropriations, 22 Stat. 861, 16 U.S.C. 559) permits Forest Service employees to make arrests for violation of laws and regulations on national forests.

Bureau of Forestry becomes Forest Service, effective July 1 (33 Stat. 861, 872-873).

1906 American Antiquities Act authorizes creation of national monuments by presidential proclamation (34 Stat. 225).

Forest Homestead Act, passed on June 11, opens agricultural lands for entry within forest reserves (34 Stat. 233); land claims filed under it were "June 11 lands."

Ten percent of receipts from forest reserves is to be returned to states or territories for benefit of public roads and schools (34 Stat. 669, 684).

1907 Forest reserves are renamed national forests; further enlargement forbidden in six western states except by act of Congress (34 Stat. 1256, 1269).

All money received by or on account of the Forest Service is required to go to Treasury (34 Stat. 1270).

1908 First Forest Service experiment station is established at Fort Valley, Arizona.

Conference of governors is held in White House.

Payments to states for schools are increased to twenty-five percent of national forest receipts (35 Stat. 251, 260).

Forest Service is reorganized by region.

1909 Western Forestry and Conservation Association is organized.

Report of National Conservation Commission is sent to Congress.

1910 President William H. Taft fires Gifford Pinchot; Henry S. Graves is named chief.

Forest Products Laboratory is dedicated at Madison, Wisconsin.

Big Blowup fires burn forests in Idaho and Montana.

1911 Weeks Law (36 Stat. 961) authorizes federal purchase of lands in watersheds with navigable streams and matching funds for state forestry agencies.

Supreme Court upholds right of Congress to create national forests and secretary of Agriculture's right to issue rules and regulations for forests and prescribe penalties for violations (220 US 506, 523).

1914 Bureau of Corporations issues report on lumber industry.

1915 Branch of Research is established in Forest Service.

Term Permit Act (16 U.S.C. 497) authorizes permits for hotels, resorts, summer homes, stores, and facilities for industrial, commercial, educational, or public uses.

1916 Chamberlain-Ferris Act (39 Stat. 218) revests federal title to unsold lands of Oregon and California Railroad Company.

National Park Service is established in Department of Interior (39 Stat. 535).

1920 Henry S. Graves resigns as chief; William B. Greeley is named to succeed him.

Capper Report, "Timber Depletion, Lumber Exports, and Concentration of Timber Ownership," is sent to Senate.

1922 Secretary of Agriculture is authorized to exchange land in national forests for private land of equal value within national forest boundaries (42 Stat. 465).

1924   Rachford Report on grazing is released.

President Calvin Coolidge convenes National Conference on Outdoor Recreation.

First wilderness area is established on Gila National Forest in New Mexico.

Clarke-McNary Act (43 Stat. 653) expands or modifies Weeks Law cooperative programs and lifts restrictions on purchase of forest lands.

1925   Forest Service begins issuing ten-year grazing permits.

1928   McNary-Woodruff Act (45 Stat. 468) authorizes $8 million to purchase land under Weeks Law.

William B. Greeley resigns as chief; Robert Y. Stuart is named as successor.

McSweeney-McNary Act (45 5tat. 699) authorizes forestry research program, including forest survey.

1930   Knutson-Vandenberg Act (45 Stat. 527) authorizes funds for reforestation of national forests and creation of revolving fund for reforestation or timber stand improvement on national forests.

President Herbert Hoover appoints Timber Conservation Board.

1932   American Forest Products Industries is incorporated as subsidiary of National Lumber Manufacturers Association.

1933   Copeland Report, "National Plan for American Forestry," is submitted to Senate.

Office of Emergency Conservation Work is established by executive order.

All national monuments are placed under Department of Interior by executive order.

National Industrial Recovery Act (48 Stat. 195) authorizes "codes of fair competition" for industries.

Soil Erosion Service is established in Department of Interior.

Chief Robert Y. Stuart dies and is succeeded by Ferdinand A. Silcox.

1934   Prairie States Forestry ("Shelterbelt") Project is started with emergency funds and administered by Forest Service.

Taylor Grazing Act (48 Stat. 1269) authorizes secretary of Interior to establish eighty million acres of grazing districts in unreserved public domain.

1935   Soil Conservation Act (49 Stat. 163) establishes Soil Conservation Service (formerly Soil Erosion Service) in Department of Agriculture.

Supreme Court invalidates National Industrial Recovery Act of 1933 (295 US 490).

Robert Marshall, Aldo Leopold, and others form Wilderness Society.

1936   Forest Service transmits "Western Range" report to Senate.

1937   Cooperative Farm Forestry (Norris-Doxey) Act (50 Stat. 188) authorizes $2.5 million annually to promote farm forestry in cooperation with states.

Civilian Conservation Corps officially succeeds Emergency Conservation Work (50 Stat. 319).

Secretary of Interior is authorized to establish sustained-yield units on revested Oregon and California Railroad lands (50 Stat. 874).

1938   Concurrent resolutions create joint Congressional Committee on Forestry (52 Stat. 1452).

1939   Chief Ferdinand A. Silcox dies; Earle H. Clapp is named acting chief.

1940   Forest Service publishes "Forest Outings" study of forest recreation.

1941   American Forest Products Industries (now American Forest Institute) launches campaign to prevent public regulation of logging on private land; Forest Farm Program is part of effort.

   "Forest Lands of United States," report by joint Congressional Committee on Forestry, is released.

1943   Lyle F. Watts is named chief; Earle H. Clapp becomes associate chief.

1944   Sustained-Yield Forest Management Act (58 Stat. 132) authorizes secretaries of Agriculture and Interior to establish cooperative sustained-yield units with private landowners and federal units.

1946   Forest Service issues first of six "Reappraisal" reports.

   General Land Office and Grazing Service are merged to form Bureau of Land Management in Department of Interior (60 Stat. 1097, 1099).

1947   Forest Pest Control Act (61 Stat. 177) makes protection of all forest lands in United States against destructive insects and disease federal policy.

1949   Supreme Court affirms constitutionality of state regulation of logging on private lands (338 US 863).

   A Sand County Almanac, by Aldo Leopold, is published.

1950   Granger-Thye Act (64 Stat. 82) broadens authority of secretary of Agriculture and adjusts range regulations.

   Cooperative Forest Management Act (64 Stat. 473) authorizes secretary of Agriculture to cooperate with state foresters in assisting private landowners.

1952   Lyle Watts retires as chief and is succeeded by Richard E. McArdle.

1954   Oregon and California Railroad lands disputed by Agriculture and Interior become national forest lands (68 Stat. 270).

1955   Multiple Use Mining Act (69 Stat. 367, 375) returns surface rights from mining claims to United States unless claim is proven valid.

1958   "Timber Resources for America's Future" is published by Forest Service.

   Outdoor Recreation Review Commission is created (72 Stat. 328).

1960   Multiple Use–Sustained Yield Act (74 Stat. 215) directs Forest Service to give equal consideration to outdoor recreation, range, timber, water, and wildlife and fish.

1962   Outdoor Recreation Review Commission's report recommends creation of outdoor recreation bureau.

   Bureau of Outdoor Recreation is created by secretarial order (497).

   Richard E. McArdle retires as chief; Edward P. Cliff is named as successor.

   Forestry Research, State Plans, Assistance (McIntire-Stennis) Act (76 Stat. 806) authorizes federal support for forestry research at land-grant colleges.

   Rachel Carson publishes *Silent Spring*.

1963   Outdoor Recreation Act (77 Stat. 49) improves coordination of eighteen federal agencies.

1963   Pinchot family donates Grey Towers to Forest Service.

1964   Wilderness Act (78 Stat. 890) sets up ten-year congressional review program for wilderness designation.

Land and Water Conservation Fund Act (78 Stat. 900) authorizes establishment of fund to support acquisition and development of state parks.

Job Corps established.

1965   National Historic Preservation Act (80 Stat. 915) requires identifying, evaluating, and protecting historical and cultural artifacts on national forests.

1966   Region 7 is eliminated; national forests in states north and east of Kentucky and Virginia are added to Region 9, renamed Eastern Region.

1968   Public Land Law Review Commission releases its report.

Wild and Scenic Rivers Act (82 Stat. 906) provides for preservation of selected rivers in their natural state.

1970   National Environmental Policy Act (83 Stat. 852) establishes Council on Environmental Quality and requires evaluation of potential environmental impacts of pending federal legislation and agency programs.

Environmental Protection Agency is created to enforce environmental standards, monitor conditions, and conduct relevant research (84 Stat. 2086); jurisdiction over pesticides is transferred from Agriculture to new agency.

Youth Conservation Corps is established.

1972   Edward P. Cliff retires as chief; John R. McGuire succeeds him.

Amendment to Federal Water Pollution Control Act (86 Stat. 816) defines standards for Environmental Protection Agency to control sources of water pollution.

Federal Environmental Pesticide Control Act (86 Stat. 973) grants federal and state agencies increased control over pesticide use.

National Forests Program is established to recruit and train volunteers to assist in Forest Service activities.

Forest Service undertakes Roadless Area Review and Evaluation (RARE) to identify potential wilderness sites.

Gene Bernardi files complaint charging Forest Service with sex discrimination.

1973   *Izaak Walton v. Butz* decision in U.S. district court rules that clearcut logging on Monongahela National Forest in West Virginia violates Organic Act of 1897.

Report of President's Advisory Panel on Timber and Environment is released.

Endangered Species Act (87 Stat. 884) establishes federal procedures for identification and protection of endangered plants and animals in their critical habitats.

1974   Forest and Rangeland Renewable Resources (Humphrey-Rarick) Planning Act (88 Stat. 476) requires long-range planning and five-year plans to ensure adequate timber supply and environmental quality.

Eastern Wilderness Act (88 Stat. 2096) extends 1964 Wilderness Act to eastern third of United States and allows wilderness designation for regenerated forests; sixteen areas are added to system.

1975   U.S. Court of Appeals for Fourth Circuit upholds 1973 *Izaak Walton v. Butz* decision banning clearcutting on national forests in West Virginia and other states in district court's jurisdiction.

1976   National Forest Management Act (90 Stat. 2949) repeals language of Organic Act that had prompted Monongahela litigation, revises Resources Planning Act planning process, and mandates greater public participation in Forest Service decision making.

  Natural Defenses Council, Inc. v. Arcata National Corporation decision holds that California Environmental Quality Act requirement for environmental impact statements applies to harvesting timber.

1977   Second roadless area review and evaluation (RARE II) begins.

  Young Adult Conservation Corps is established to provide disadvantaged youths with work on conservation projects on public lands.

1978   Endangered American Wilderness Act (92 Stat. 40) designates largest single addition in wilderness system, totaling 1.3 million acres in ten western states.

  Sealed bids provision of National Forest Management Act is repealed.

  Amendment to Endangered Species Act allows exemptions in specific cases determined by committee of cabinet members and other high-ranking officials ("God Squad").

  Forest and Rangeland Renewable Resources Research Act of 1978 (92 Stat. 353) authorizes renewable resources research on national forest and rangelands, including research on fish and wildlife and their habitats.

  Cooperative Forestry Assistance Act (92 Stat. 365) establishes cooperative federal, state, and local forest stewardship program for management of nonfederal forest lands.

1979   John R. McGuire retires as chief; R. Max Peterson is named successor.

  Forest Service completes RARE II, basis for congressional designation of national forest wilderness areas during 1980s.

  Forest Service signs consent decree and agrees to increase hiring and promotion of women.

1980   Mount St. Helens erupts on the Gifford Pinchot National Forest.

1985   Forest Service links all agency personnel through Data General computer system known as Forest Level Information Processing System (FLIPS), becoming first government agency to use electronic mail.

1987   R. Max Peterson retires as chief; F. Dale Robertson succeeds him.

1988   Fires in Yellowstone National Park and adjacent national forest lands pressure Forest Service and National Park Service to change their "let-burn" policy.

1989   Congress votes to halt timbering in Alaska's Tongass National Forest, last undisturbed temperate rain forest in United States.

1990   Food, Agriculture, Conservation and Trade Act (1990 Farm Bill) authorizes incentive and assistance programs to conserve and maintain existing forest lands under Title XII, or Forest Stewardship Act (104 Stat. 3521), including Forest Legacy Program, Forest Stewardship Program, Forestry Incentives Program, and Stewardship Incentives Program.

  Robertson announces New Perspectives management approach.

  Interagency Scientific Committee is appointed to develop conservation strategy in Pacific Northwest.

1990 Global Climate Change Prevention Act (104 Stat. 4058) authorizes creation of Office of International Forestry and urban forestry demonstration projects.

U.S. Fish and Wildlife Service lists northern spotted owl for protection.

Tongass Timber Reform Act (104 Stat. 4426) repeals Alaska National Interest Lands Conservation Act of 1980, leading to closing of pulp mills in Sitka and Ketchikan in 1993 and 1997 respectively and termination of long-term timber supply contracts between Forest Service and the mills.

1992 President George H. W. Bush and Chief Robertson announce that clearcutting will no longer be dominant silviculture method and make ecosystem management official policy at Earth Summit in Rio de Janiero.

1993 President Clinton holds "Timber Summit" in Portland, Oregon, and orders Forest Ecosystem Management Assessment Team to produce management plan in ninety days.

F. Dale Robertson retires as chief; Jack Ward Thomas is appointed to succeed him.

Interior Columbia Basin Ecosystem Management Project (ICBEMP) is created.

1994 Smokey Bear celebrates 50th anniversary one month after fourteen forest firefighters are killed in South Canyon Fire on Storm King Mountain.

Northwest Forest Plan is signed by President Clinton and upheld in court.

1995 Rescission Act (109 Stat. 194) authorizes funds for Emergency Salvage Timber Sale Program ("salvage rider").

1996 Seventh American Forest Congress, held in Washington, calls on diverse participants to develop common vision for American forests.

1997 Jack Ward Thomas retires as chief; Michael P. Dombeck succeeds him.

1998 Dombeck introduces Natural Resource Agenda.

1999 Dombeck announces plans to suspend new road construction in most inventoried roadless areas.

2000 Fire season consumes more than 4.8 million acres of forest and $1.6 billion in fire suppression costs.

Forest Service issues National Fire Plan to address fuel buildup on national forests.

2001 Michael P. Dombeck retires as chief; Dale N. Bosworth is named his successor.

Forest Service and National Interagency Fire Center send incident management teams to help in rescue and recovery efforts when World Trade Center and Pentagon are attacked.

2003 Bosworth introduces his Four Threats agenda.

Healthy Forests Restoration Act (117 Stat. 1887) is passed after national attention focuses on wildfire.

2005 Forest Service celebrates its centennial.

Appendix B
Biographical Sketches of the Chiefs

Gifford Pinchot

Gifford Pinchot (1898–1910) served as chief of the U.S. Division of Forestry from 1898 to 1905, and then the Forest Service from its establishment in 1905 to 1910. Pinchot graduated from Yale in 1889, a time when there were no forestry schools in the United States; he went to Europe to study forestry. He returned in 1890 and implemented forest management plans for private landholders in North Carolina and New York while working for passage of federal legislation establishing management on the federal forest reserves (later called national forests). He was the first American-born forester appointed division chief. His division worked closely with the Department of the Interior to introduce forestry on public lands and with private landholders to do the same on private lands, which gave his men much needed field experience. Meanwhile, Pinchot gathered congressional support for the transfer of the forest reserves from Interior to his division in the Agriculture Department. When the transfer came in 1905, Pinchot and the new Forest Service took the lead in making conservation a national objective. With the help of his subordinates, he began establishing the Forest Service's administrative structure. His zeal for conservation led to a fallout with President Taft, who fired him for insubordination in 1910. He also established the Society of American Foresters and, with his family, founded the Yale Forest School, both in 1900. In doing so, he laid the foundation for professional forestry in the United States.

Henry S. Graves

Henry S. Graves (1910–1920) graduated from Yale in 1892 and, on the advice of his college friend Gifford Pinchot, studied forestry in Germany before beginning working with Pinchot on various forestry jobs. When Pinchot became head of the Division of Forestry, Graves served as his assistant chief. In 1900, Graves was appointed dean of forestry at the new Yale Forest School and quickly built it into a leading educational institution, and he was one of the original seven members of the Society of American Foresters. He was selected to take over the five-year-old Forest Service after Pinchot was fired in 1910. Graves restored relations with the Department of the Interior and revived a Forest Service demoralized by Pinchot's dismissal. He also fought off state and private interests that wanted the forests returned to state or local control. In 1917, Graves was commissioned a major in the U.S. Army and served in the Tenth Engineers (Forestry) and later the Twentieth Engineers. His tenure as chief was characterized by a stabilization of the national forests, purchase of new national forests in the East, and a strengthening of the scientific foundations of forestry. His greatest contribution was the successful launching of a national forest policy for the United States.

William B. Greeley (1920–1928) graduated in 1901 from the University of California and three years later from the Yale Forest School, where Dean Henry Graves marked him as a rising star. After starting with the Bureau of Forestry in 1904, he quickly was promoted through the ranks. His work during the Big Blowup fire in 1910 was rewarded with a promotion to the Washington office as assistant chief in charge of silviculture. During the Great War, Greeley led the Army's Twentieth Engineers (Forestry), ending the war as a lieutenant colonel. As chief, Greeley put into practice the national forest policy he helped craft under Chief Graves. Greeley made cooperation with private, state, and other federal agencies a standard feature of Forest Service management, struggled with the National Park Service over numerous issues, and "blocked up" the national forests through the exchange or purchase of lands inside or near the forest boundaries. Passage of the Clarke-McNary Act of 1924, which expanded or modified many Weeks Law cooperative agreements, and the McSweeney-McNary Act of 1928, which promoted forest research, were two highlights of his administration.

William B. Greeley

Robert Y. Stuart (1928–1933) graduated from Dickinson College in 1903 and then entered Yale Forest School, where he received a master of forestry degree in 1906. He worked in national forests in the West for six years before coming to the Washington office in 1912. In fall 1917, Stuart was commissioned a captain in the Army's Twentieth Engineers (Forestry) in France. Having returned briefly to the Forest Service after the Great War, he resigned to work in forestry for Pennsylvania under Governor Gifford Pinchot. He began a program to buy land for the state forest system, established a statewide system of forest fire lookouts, and started a forest nursery system. He returned to the Forest Service in 1927 and was appointed chief shortly thereafter. During the Great Depression, Stuart led the Forest Service in creating job opportunities for the unemployed on the national forests; the road system was one result. During his term, the McSweeney-McNary Act of 1928 promoted forest research, and the Knutson-Vandenburg Act of 1930 expanded tree planting on the national forests. He died from a fall out of his office window.

Robert Y. Stuart

Ferdinand Augustus "Gus" Silcox (1933–1939) graduated from the College of Charleston, South Carolina, in 1903, with honors in chemistry and sociology. He received a master's degree in forestry from the Yale Forest School in 1905 and served with the Forest Service in the northern Rockies after graduation. He was there during the Big Blowup of 1910. Silcox joined the Army's forest engineers division in 1917 as a captain and handled labor problems at the shipyards in the Puget Sound and Columbia River districts. After the war, Silcox worked in the private sector for eleven years as a director of industrial relations before being appointed chief. An able administrator, during the Great Depression he directed the Civilian Conservation Corps and Works Projects Administration projects in the national forests. He also pushed for control of timber-cutting operations on private lands. He died of a heart attack while in office.

Ferdinand Augustus Silcox

Earle H. Clapp

Lyle Watts

Richard E. McArdle

Earle H. Clapp (Acting) (1939–1943) received his B.A. in forestry in 1905 from the University of Michigan. He first started to work for the Forest Service on the Medicine Bow Forest Reserve as a timber surveyor. In 1906, he worked on several forest reserves (now national forests) to develop techniques for determining minimum prices for timber. The following year, Clapp was appointed chief of timber sales in the national office. In 1909, he worked in the national forests in the Southwest and then in 1915 was made the chief of the new Forest Service Branch of Research. Among his accomplishments as research chief was the Copeland Report in 1933. He was appointed associate chief in 1935 and then acting chief in 1939 after Chief Silcox died. Clapp was never officially named chief of the agency and served in an acting capacity until Lyle Watts was appointed. Clapp met the need for forest experts to help in the aftermath of the disastrous New England hurricane of September 1938. Timber operations salvaged seven hundred million board feet of timber to prevent insect and disease infestations and prevent fires. He also thwarted transfer of the Forest Service from Agriculture to Interior and helped prepare for World War II and the attendant demand for forest products.

Lyle Watts (1943–1952) earned a B.S. in 1913 and later his M.S.F. in forestry from Iowa State College. He began working for the Forest Service while in college and joined the agency in 1913. He quickly rose through the ranks to serve as supervisor on three national forests in Idaho between 1918 and 1926 and then served as assistant chief for forest management in Region 4 from 1926 to 1928. He left for fifteen months to organize the department of forestry at Utah State Agricultural College (now Utah State University). Back at the Forest Service, he was a senior silviculturist and directed watershed studies at the experiment station in Ogden, Utah. After serving as director of the experiment station from 1931 to 1936 at Missoula, Montana, he was appointed regional forester for the North Central region and, after three years, took the same position in the Pacific Northwest for three more years before being named chief in 1943. High demand for timber during his term led to increased calls for federal control over cutting on private lands. He also encouraged the Forest Service to hire university forestry graduates after the war to help develop and manage the forests as the agency embarked on a period of intensive resource management.

Richard E. McArdle (1952–1962) earned a B.S. degree in 1923 and his M.S. in 1924 from the University of Michigan, where he studied under Filibert Roth, who had led the Interior Department's old forestry division. McArdle then joined the Forest Service as a silviculturist, worked for three years at the Pacific Northwest Forest and Range Experiment Station in Oregon before returning to Michigan to earn his doctorate in botany, and then worked for the agency in research. After serving as director of two research stations, he went to Washington in 1944 to serve as assistant chief for State and Private Forestry cooperative programs. Named chief in

1952, he was the first to hold a Ph.D. and to have been a researcher. He ended debate over regulation of timber-harvesting practices on private lands, which improved relations with the timber industry. The Multiple Use–Sustained Yield Act of 1960 established policy for the broad development and administration of the national forests in the public interest and led to the increasingly intensive management of the national forests. During his tenure, the Forest Service accepted responsibility for the national grasslands.

Edward P. Cliff

Edward P. Cliff (1962–1972) graduated with a forestry degree from Utah State College in 1931. He started with the Forest Service the same year on the Wenatchee National Forest in Washington and stayed in the Pacific Northwest until 1944, when he went to the Washington office. Two years later, he was assigned to the Intermountain Region in charge of range and wildlife and became regional forester for the Rocky Mountain Region in 1950. In 1952, he returned to Washington as assistant chief of the Forest Service and then was appointed chief in 1962. Public interest in the management of the national forests, as well as demands for numerous forest resources, expanded quickly during his administration. Legislation passed while Cliff was in office included the Wilderness Act, the first two versions of the Endangered Species Act, and the National Environmental Policy Act.

John R. McGuire (1972–1979) graduated with a degree in forestry from the University of Minnesota. He briefly worked for the Forest Service before his interest in forest research led him to earn his M.F. degree from Yale University in 1941. He worked at the Forest Service research facility on campus. After the war, when he served in the Army Corps of Engineers in the Pacific Theater, he took a research position at the Forest Service's Northeastern Forest Experiment Station in New Haven, Connecticut. He moved in 1950 to a research station at Upper Darby, Pennsylvania, while completing his M.A. in economics at the University of Pennsylvania. In 1962, McGuire became director of the Pacific Southwest Forest and Range Experiment Station in Berkeley, California. He moved to the national office in 1967 and became chief in 1972. The Endangered Species Act of 1973, the Resources Planning Act, and the National Forest Management Act were passed while McGuire was chief. He faced protest over the use of chemical pesticides, the practice of clearcutting, and the roadless area reviews.

John R. McGuire

R. Max Peterson (1979–1987), the first nonforester named to head the Forest Service, received a bachelor's degree in 1949 in civil engineering from the University of Missouri after serving in naval aviation during World War II. He worked as an engineer in the Plumas National Forest and several other national forests in California. He earned a master's degree in public administration in 1959 from Harvard University. In 1961, he was assigned to the agency's national office, where he remained until 1966, when he was transferred back to California to

R. Max Peterson

F. Dale Robertson

Jack Ward Thomas

Michael P. Dombeck

serve as regional engineer. He was made regional forester for the Southern Region in 1972, then returned to Washington in 1974. Peterson served during a time of increasing turmoil and criticism of the Forest Service as the agency tried to implement the National Forest Management Act and other legislation while coping with a reduced budget and payroll; debate raged over "wise use" and environmental concerns.

F. Dale Robertson (1987–1993) joined the Forest Service in 1961 as a ranger after receiving a degree in forestry from the University of Arkansas. His early assignments were in the South. After moving to the Washington office in 1968 to work as a management analyst, he completed a master's degree in public administration from American University in Washington, D.C., in 1970. Shortly afterward, he was reassigned to the Pacific Northwest. In 1980, he returned to Washington as associate deputy chief for programs and legislation; integrating regional and individual forest plans into the National Forest System, as mandated by the Resources Planning Act, was among his responsibilities. In 1982, he became Peterson's associate chief and then chief five years later. He was the first ranger to rise to the rank of chief. Robertson tackled employment issues, introduced ecosystem management, and dealt with the highly contentious northern spotted owl issue.

Jack Ward Thomas (1993–1996) received a B.S. degree in wildlife management from Texas A&M University, then his M.S. in wildlife ecology from West Virginia University in 1969 and a Ph.D. in forestry (natural resources planning) from the University of Massachusetts in 1972. He worked for the Texas Game and Fish Commission from 1956 to 1966 before joining the Forest Service in Morgantown, West Virginia, as a research wildlife biologist. In 1969, he moved to the Urban Forestry and Wildlife Research Unit at Amhurst, Massachusetts. Five years later, he became the chief research wildlife biologist and project leader at the Blue Mountains Research Lab in La Grande, Oregon, where he became involved in ecosystem management. During his administration, Thomas attempted to revive a demoralized agency while trying to implement ecosystem management. He also worked closely with Michael Dombeck, acting director of the Bureau of Land Management, to better coordinate the land management efforts of the two agencies.

Michael P. Dombeck (1997–2001) earned undergraduate and graduate degrees in biological sciences and education from the University of Wisconsin–Stevens Point and the University of Minnesota and earned his doctorate in fisheries biology from Iowa State University. He joined the Forest Service in 1978 and spent almost 12 years with the agency as a fisheries biologist. Beginning in 1989, he served as science adviser and special assistant to the director of the Bureau of Land Management and was named acting director in 1994. He became Forest Service chief in 1997. He is the only person to have served as the head of the two largest land management agencies. Dombeck's Natural Resource Agenda, which included new roadless area rules, helped reinvigorate the agency by providing long-term, tangible goals for ecosystem management.

Dale Bosworth (2001–) received a B.S. in forestry from the University of Idaho in 1966 and began working for the agency on the St. Joe National Forest (now part of the Idaho Panhandle National Forest) in Idaho. Several positions later, Bosworth moved to the Flathead National Forest as the planning staff officer before becoming the deputy forest supervisor there. He then served as the assistant director for land management planning for Region 1. In 1986, Bosworth was named forest supervisor of the Wasatch-Cache National Forest in Utah. Four years later, he became deputy director of forest management in the Forest Service national headquarters, where he served until 1992, when he became deputy regional forester for Region 5 for two years. From 1994 until his appointment as chief in 2001, he served as regional forester for Region 4 and then Region 1. Bosworth's Four Threats agenda has guided his administration's efforts, which have focused on ecological restoration.

Dale Bosworth

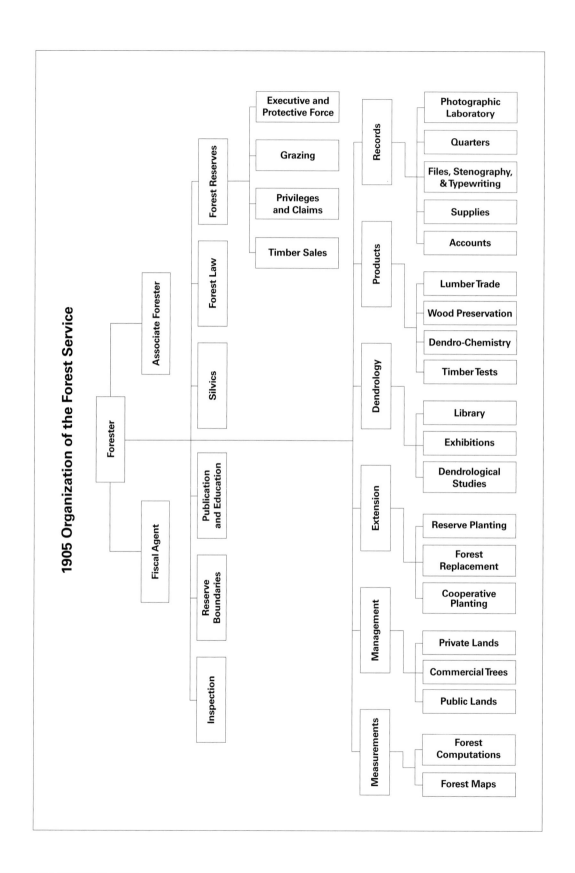

1905 Organization of the Forest Service

- **Forester**
  - **Associate Forester**
    - **Forest Reserves**
      - Executive and Protective Force
      - Grazing
      - Privileges and Claims
      - Timber Sales
    - **Forest Law**
    - **Silvics**
  - **Fiscal Agent**
    - **Publication and Education**
    - **Reserve Boundaries**
    - **Inspection**
  - **Records**
    - Photographic Laboratory
    - Quarters
    - Files, Stenography, & Typewriting
    - Supplies
    - Accounts
  - **Products**
    - Lumber Trade
    - Wood Preservation
    - Dendro-Chemistry
    - Timber Tests
  - **Dendrology**
    - Library
    - Exhibitions
    - Dendrological Studies
  - **Extension**
    - Reserve Planting
    - Forest Replacement
    - Cooperative Planting
  - **Management**
    - Private Lands
    - Commercial Trees
    - Public Lands
  - **Measurements**
    - Forest Computations
    - Forest Maps

1907 Organization of the Forest Service

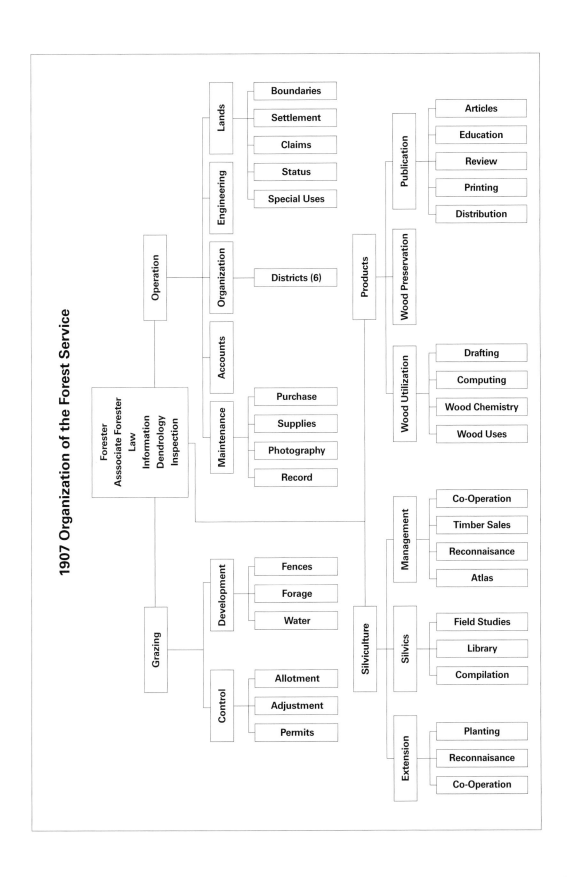

Forester
Asssociate Forester
Law
Information
Dendrology
Inspection

Operation
- **Lands**
  - Boundaries
  - Settlement
  - Claims
  - Status
  - Special Uses
- **Engineering**
- **Organization**
  - Districts (6)
- **Accounts**
- **Maintenance**
  - Purchase
  - Supplies
  - Photography
  - Record

Grazing
- **Development**
  - Fences
  - Forage
  - Water
- **Control**
  - Allotment
  - Adjustment
  - Permits

Products
- **Publication**
  - Articles
  - Education
  - Review
  - Printing
  - Distribution
- **Wood Preservation**
- **Wood Utilization**
  - Drafting
  - Computing
  - Wood Chemistry
  - Wood Uses

Silviculture
- **Management**
  - Co-Operation
  - Timber Sales
  - Reconnaisance
  - Atlas
- **Silvics**
  - Field Studies
  - Library
  - Compilation
- **Extension**
  - Planting
  - Reconnaisance
  - Co-Operation

1954 Organization of the Forest Service

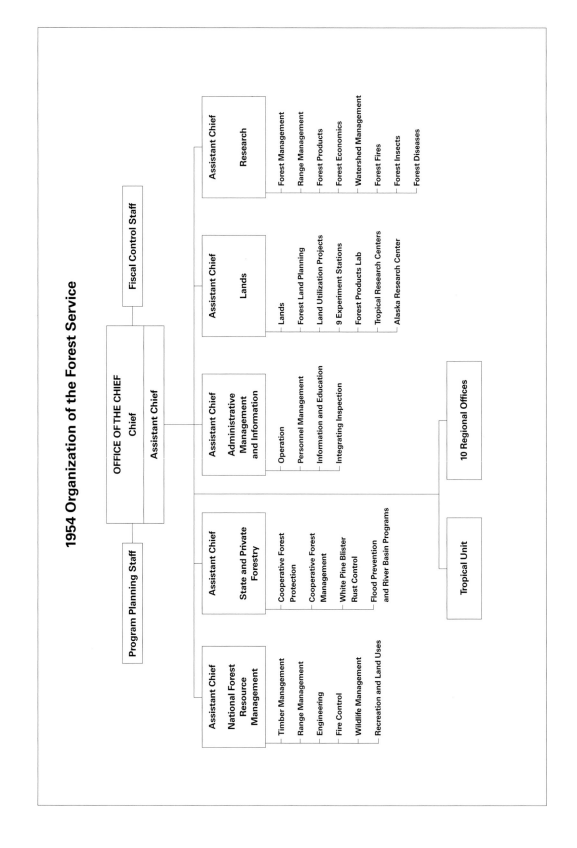

OFFICE OF THE CHIEF
Chief
Assistant Chief

Program Planning Staff

Fiscal Control Staff

Assistant Chief
National Forest Resource Management
- Timber Management
- Range Management
- Engineering
- Fire Control
- Wildlife Management
- Recreation and Land Uses

Assistant Chief
State and Private Forestry
- Cooperative Forest Protection
- Cooperative Forest Management
- White Pine Blister Rust Control
- Flood Prevention and River Basin Programs

Assistant Chief
Administrative Management and Information
- Operation
- Personnel Management
- Information and Education
- Integrating Inspection

Assistant Chief
Lands
- Lands
- Forest Land Planning
- Land Utilization Projects
- 9 Experiment Stations
- Forest Products Lab
- Tropical Research Centers
- Alaska Research Center

Assistant Chief
Research
- Forest Management
- Range Management
- Forest Products
- Forest Economics
- Watershed Management
- Forest Fires
- Forest Insects
- Forest Diseases

Tropical Unit

10 Regional Offices

2003 Organization of the Forest Service

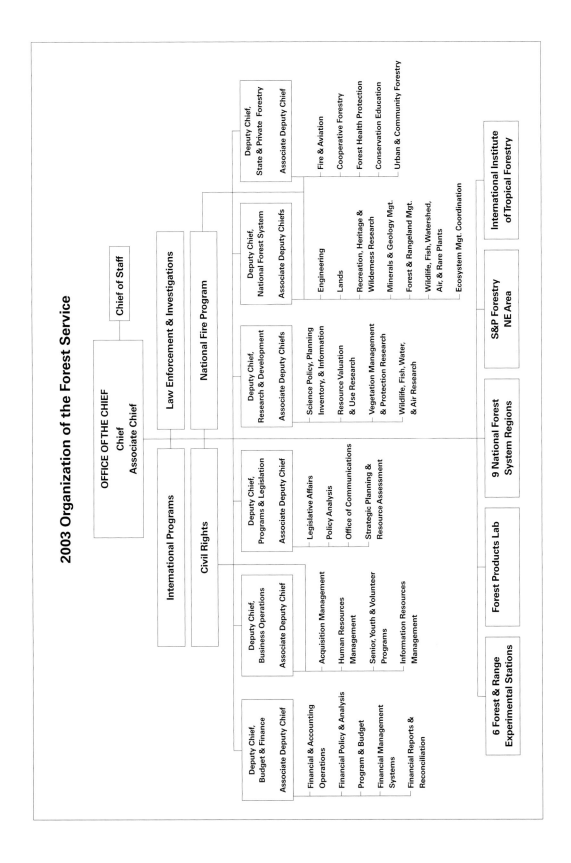

OFFICE OF THE CHIEF
Chief
Associate Chief

Chief of Staff

International Programs

Civil Rights

Law Enforcement & Investigations

National Fire Program

Deputy Chief, Budget & Finance
Associate Deputy Chief

- Financial & Accounting Operations
- Financial Policy & Analysis
- Program & Budget
- Financial Management Systems
- Financial Reports & Reconciliation

Deputy Chief, Business Operations
Associate Deputy Chief

- Acquisition Management
- Human Resources Management
- Senior, Youth & Volunteer Programs
- Information Resources Management

Deputy Chief, Programs & Legislation
Associate Deputy Chief

- Legislative Affairs
- Policy Analysis
- Office of Communications
- Strategic Planning & Resource Assessment

Deputy Chief, Research & Development
Associate Deputy Chiefs

- Science Policy, Planning Inventory, & Information
- Resource Valuation & Use Research
- Vegetation Management & Protection Research
- Wildlife, Fish, Water, & Air Research

Deputy Chief, National Forest System
Associate Deputy Chiefs

- Engineering
- Lands
- Recreation, Heritage & Wilderness Research
- Minerals & Geology Mgt.
- Forest & Rangeland Mgt.
- Wildlife, Fish, Watershed, Air, & Rare Plants
- Ecosystem Mgt. Coordination

Deputy Chief, State & Private Forestry
Associate Deputy Chief

- Fire & Aviation
- Cooperative Forestry
- Forest Health Protection
- Conservation Education
- Urban & Community Forestry

6 Forest & Range Experimental Stations

Forest Products Lab

9 National Forest System Regions

S&P Forestry NE Area

International Institute of Tropical Forestry

Further Reading

Any number of books discuss conservation and the beginning and current operations of the Forest Service. A short list of readings should include:

Catton, Theodore, and Lisa Mighetto. *The Fish and Wildlife Job on the National Forests: A Century of Game and Fish Conservation, Habitat Protection, and Ecosystem Management.* Missoula, MT: Historical Research Associates for USDA Forest Service, 1999.

Cox, Thomas R., Robert S. Maxwell, Phillip Drennon Thomas, and Joseph J. Malone. *This Well-Wooded Land: Americans and Their Forests from Colonial Times to the Present.* Lincoln: University of Nebraska Press, 1985.

Clary, David A. *Timber and the Forest Service.* Lawrence: University Press of Kansas, 1986.

Dana, Samuel Trask, and Sally K. Fairfax. *Forest and Range Policy: Its Development in the United States,* 2nd ed. New York: McGraw-Hill Book Company, 1956, 1980.

Dombeck, Michael P., Christopher A. Wood, and Jack E. Williams. *From Conquest to Conservation: Our Public Lands Legacy.* Washington, DC: Island Press, 2003.

Fedkiw, John. *Managing Multiple Uses on National Forests, 1905–1995: A 90-Year Learning Experience and It Isn't Finished Yet.* FS-628. Washington, DC: USDA Forest Service, 1998.

Frome, Michael. *The Forest Service.* 2nd ed. Boulder, CO: Westview Press, 1984.

Hays, Samuel P. *Conservation and the Gospel of Efficiency: The Progressive Conservation Movement, 1890–1920.* Cambridge, MA: Harvard University Press, 1959.

Hirt, Paul W. *A Conspiracy of Optimism: Management of the National Forests Since World War Two.* Lincoln: University of Nebraska Press, 1994.

Meine, Curt. *Aldo Leopold: His Life and Work.* Madison: University of Wisconsin Press, 1988.

Miller, Char. *Gifford Pinchot and the Making of Modern Environmentalism.* Washington, DC: Island Press, 2001.

Pinchot, Gifford. *Breaking New Ground.* Commemorative Edition, with an Introduction by Char Miller and V. Alaric Sample. Washington, DC: Island Press, 1998.

Steen, Harold K. *The U.S. Forest Service: A History.* Seattle: University of Washington Press, 1976; Centennial Edition, 2004, with a new preface by the author.

Steen, Harold K. (ed.) *The Origins of the National Forests: A Centennial Symposium.* Durham, NC: Forest History Society, 1992.

Steen, Harold K. (ed.) *Forest and Wildlife Science in America: A History.* Durham, NC: Forest History Society, 1999.

Thomas, Jack Ward. *The Journals of a Forest Service Chief,* ed. by Harold K. Steen. Durham, NC: Forest History Society and University of Washington Press, 2004.

Williams, Gerald W. *The USDA Forest Service: The First Century.* FS-650. Washington, DC: USDA Forest Service, 2005.

Williams, Michael. *Americans and Their Forests: A Historical Geography.* Cambridge: Cambridge University Press, 1989.

Endnotes

Chapter One: Origins of the Forest Service

1. Michael Williams, *Americans and Their Forests: A Historical Geography* (Cambridge: Cambridge University Press, 1989), 118.

2. For details on other uses, such as paper pulp and railroad ties, and the effects of changing technology on lumber demands, see Williams, *Americans and Their Forests*, 53–192.

3. Samuel Trask Dana and Sally K. Fairfax, *Forest and Range Policy: Its Development in the United States*, 2nd ed., (New York: McGraw-Hill Book Company, 1956, 1980), 37–38.

4. Williams, *Americans and Their Forests*, 1–189, examines American forest history from before European settlement to 1859 and specifically discusses attitudes toward the forest prior to the mid-nineteenth century (14–18). Roderick Nash, in *Wilderness and the American Mind*, 3rd ed. (New Haven: Yale University Press, 1982), 1–107, discusses forests in American thought in the broader context of wilderness.

5. See Donald Pisani, "Forests and Conservation, 1865–1890," in *American Forests: Nature, Culture, and Politics*, edited by Char Miller (Lawrence: University Press of Kansas, 1997), 22–25, for a discussion on the intellectual debate and origins of the forest preservation movement found in the literature of the period.

6. Gifford Pinchot, *Breaking New Ground*, Commemorative Edition, with an Introduction by Char Miller and V. Alaric Sample (Washington, DC: Island Press, 1998), 13.

7. George Perkins Marsh, *Man and Nature: Or, Physical Geography as Modified by Human Action*, edited and annotated by David Lowenthal (Cambridge: Belknap Press of Harvard University, 1965). Dana and Fairfax, *Forest and Range Policy*, 38–39, discuss the popularity of the book when published. In 2004, an entire issue of *Environment and History* focused on Marsh's work and its impact. See *Environment and History* 10(2) (May 2004), "The Nature of G.P. Marsh: Tradition and Historical Judgement," edited by Marcus Hall.

8. In his memoirs, Gifford Pinchot mentions having read Marsh while in college. Pinchot, *Breaking New Ground*, 4. Philip Shabecoff asserts that naturalist John Muir was also influenced by Marsh in *A Fierce Green Fire: The American Environmental Movement* (New York: Hill and Wang, 1993), 58. See also Donald Pisani, "Forests and Reclamation, 1891–1911," in *The Origins of the National Forests: A Centennial Symposium*, edited by Harold K. Steen (Durham, NC: Forest History Society, 1992), 238.

9. Marsh, *Man and Nature*, 230.

10. Marsh, *Man and Nature*, 45–46.

11. Marsh, *Man and Nature*, xxi–iii; and Shabecoff, *Green Fire*, 55–59.

12. Howard S. Miller, "The Political Economy of Science," in *Nineteenth-Century American Science: A Reappraisal*, edited by George H. Daniels (Evanston, IL: Northwestern University Press, 1972), 97.

13. Marsh, *Man and Nature*, 233–35, 259–63; and Joseph M. Petulla, *American Environmental History: The Exploitation and Conservation of Natural Resources* (San Francisco: Boyd & Fraser Publishing, 1977), 221.

14. Thomas R. Cox, Robert S. Maxwell, Phillip Drennon Thomas, and Joseph J. Malone, *This Well-Wooded Land: Americans and Their Forests from Colonial Times to the Present* (Lincoln: University of Nebraska Press, 1985), 147–48.

15. Frank E. Smith, *The Politics of Conservation* (New York: Pantheon Books, 1966), 53. Lawmakers initially appeared prescient when the Great Plains experienced above-average rainfall beginning in the 1870s and continuing into the 1890s. Unfortunately, the increased rainfall resulted not from their legislative efforts or from planting trees, but from the region's normal climatic cycle. Cox et al., *Well-Wooded Land*, 148. The legislators were mistaken in their assumptions about environmental conditions and the applicability of farming techniques that worked in the more humid eastern section of the United States. See Dana and Fairfax, *Forest and Range Policy*, 21–32, for more about the laws. Harold K. Steen offers a concise discussion of the problems of disposition of timberland in *The U.S. Forest Service: A History* (Seattle: University of Washington Press, 1976; Centennial Edition, 2004, with a new preface by the author), 7.

16. Michael Frome, *The Forest Service*, 2nd ed. (Boulder, CO: Westview Press, 1984), 13; and Williams, *Americans and Their Forests*, 232–33, 449.

17. See Petulla, *Environmental History*, 222–25, for details on events during the 1870s and 1880s. For more on the establishment of the two organizations, see William G. Robbins, *American Forestry: A History of National, State, and Private Cooperation* (Lincoln: University of Nebraska Press, 1985), 3–4; and Herbert Donald Kirkland, II, "The American Forests, 1864–1898: A Trend Towards Conservation" (Ph.D. dissertation, Florida State University, 1971), 94–97. According to Kirkland, efforts began with the American Forestry Congress. The Congress was a separate entity from the first American

Forestry Association. Dr. John A. Warder, a physician interested in forestry, had formed AFA in 1875. The organizers of the congress invited members of the American Forestry Association to their Cincinnati meeting in 1882. Though there had been no plans to unite the two groups, Warder's stature in both groups led to their merger later that year, after Franklin Hough and others convinced him that the nascent American forestry movement could not support two competing organizations. In 1889, it formally changed its name to the American Forestry Association; more recently it has become American Forests.

[18] For more on Hough's career in forestry, see Steen, *Forest Service*, 9–21; and Frank J. Harmon, "Remembering Franklin B. Hough," *American Forests* 83 (January 1977): 34–37, 52–54.

[19] Marsh, *Man and Nature*, xxii.

[20] Congress, Senate, *Message from the President of the United States communicating information in relation to the cultivation of timber and the preservation of forests*, 43rd Cong., 1st sess., Ex. Doc. 28, (19 February 1874).

[21] Steen, *Forest Service*, 10.

[22] The manner in which Hough's position came into being—via an amendment to another bill—would be repeated two more times, first with the passage of the Forest Reserve Act (1891) and again with the Forest Management Act (1897). Those two laws set up the National Forest System and directed how the government would manage the forests. In fact, not until passage of the Multiple Use–Sustained Yield Act in 1960 would Congress pass a law that directly affected forest management as a standalone bill.

[23] Steen, *Forest Service*, 14.

[24] Filibert Roth, who led the Department of the Interior's Division of Forestry for a brief period at the beginning of the twentieth century, complained that the lack of funding hindered his agency's work. Filibert Roth, "Administration of U.S. Forest Reserves, Part 1," *Forestry and Irrigation* VIII(5) (May 1902): 191–92.

[25] Steen, *Forest Service*, 143, 170.

[26] Commission of Agriculture. *Report upon Forestry*, Vols. 1–3, by Franklin B. Hough (Washington, DC: Government Printing Office, 1878, 1880, 1882); Steen, *Forest Service*, 15–18; and Frome, *Forest Service*, 14.

[27] Cox et al., *Well-Wooded Land*, 181.

[28] Nathaniel Egleston, though a member of the American Forestry Association, had been appointed chief as a political favor. Gifford Pinchot characterized him as "one of those failures in life whom the spoils system is constantly catapulting into responsible positions" and

his administration as "three years of innocuous desuetude." The passage reveals not only Pinchot's disgust for the unreformed civil service but also his disdain for untrained men interloping in forestry affairs. See Pinchot, *Breaking New Ground*, 135. Steen, *Forest Service*, 20–21, notes that what little documentary evidence exists supports the substance of the description.

[29] S. B. Sutton, *Charles Sprague Sargent and the Arnold Arboretum* (Cambridge: Harvard University Press, 1970), 80.

[30] Stephen Fox, *The American Conservation Movement: John Muir and His Legacy* (Madison: University of Wisconsin Press, 1981), 109.

[31] Sutton, *Sargent*, 12–73, *passim*.

[32] Steen, *Forest Service*, 83; and Yale Forest School, *Biographical Record of the Graduates and Former Students of the Yale Forest School, with Introductory Papers on Yale in the Forestry Movement and the History of the Yale Forest School* (New Haven: Yale Forest School, 1913), 19.

[33] Sutton, *Sargent*, 79. Brewer's work was not published until 1874.

[34] Williams, *Americans and Their Forests*, 23. Williams discusses the development of knowledge about American forests in the context of mapping them in *Americans and Their Forests*, 22–30. William Bartram and his son John conducted the first comprehensive examination of North American flora in the deep coastal South in the last quarter of the eighteenth century. Another father-and-son team, the French botanists André Michaux and Francois André Michaux, his son, published on individual species in the early 1800s, and Thomas Nuttall published a supplement to their work in 1818. Many of their observations, and those of others, about the interconnectedness and dynamism of natural landscapes finally found an interested public when Marsh used their research in his book. See a discussion of Marsh's intellectual antecedents in Richard W. Judd, "George Perkins Marsh: The Times and Their Man," *Environment and History* 10(2) (May 2004): 169–90.

[35] Pisani, "Forests and Conservation," 15–16.

[36] Pisani, "Forests and Conservation," 23; and Fox, *American Conservation*, 109.

[37] Congress, House. *Report on the Forests of North America (exclusive of Mexico)*, by Charles S. Sargent, 47th Cong., 2nd sess., HR Miscellaneous Doc. 42, Part 9 (1884), 493.

[38] Sargent, *Forests of North America*, especially 490–95. For a summary of Hough's work, see Congress, House, Forestry. *Message from the President of the United States. Transmitting a special report upon the subject of forestry by the Commissioner of Agriculture*, 45th Cong., 2nd sess., HR Exec. Doc. 24 (13 December 1877).

39 Jenks Cameron, *The Development of Governmental Forest Control in the United States* (Baltimore: Johns Hopkins University Press, 1928), 196.

40 James G. Lewis, "'Trained by Americans in American Ways': The Establishment of Forestry Education in the United States, 1885–1911" (Ph.D. dissertation, Florida State University, 2001), 28.

41 Fernow served in the Prussian Forest Service for seven years and completed his forestry training at the Forest Academy at Muenden. As part of his training, Fernow entered the forest service and worked for one year before passing admission examinations. Of those seven years, Fernow spent one year in the German army fighting in the Franco-Prussian War and, following his return home, another year at the University of Konigsberg studying law.

42 Steen, *Forest Service*, 24; and Andrew Denny Rodgers III, *Bernhard Eduard Fernow: A Story of North American Forestry* (Princeton: Princeton University Press, 1951), 14. For more on Fernow's background and education, see Rodgers, *Fernow*, 14–17.

43 Rodgers, *Fernow*, 25–26. For more on the iron industry and its voracious consumption of wood, see Williams, *Americans and Their Forests*, 104–110.

44 Robbins, *American Forestry*, 4–5.

45 For more on this, see Steen, *Forest Service*, 17–21.

46 A. Hunter Dupree, *Science in the Federal Government: A History of Policies and Activities* (Baltimore: Johns Hopkins University Press, 1986), 211.

47 Steen, *Forest Service*, 20–21, 23–24.

48 Char Miller, "Wooden Politics: Bernhard Fernow and the Quest for a National Forest Policy, 1876–1898," in *The Origins of the National Forests: A Centennial Symposium*, edited by Harold K. Steen (Durham, NC: Forest History Society, 1992), 287.

49 Steen, *Forest Service*, 37. For more on Fernow's tenure in office, see Steen, *Forest Service*, 37–46; Rodgers, *Fernow*, 196–252; and Miller, "Wooden Politics," 287–300.

50 Barbara McMartin, *The Great Forest of the Adirondacks* (Utica, NY: North Country Books, 1994), 215.

51 For more on Fernow's frustrations while working for the government, see Miller, "Wooden Politics," 295; and Steen, *Forest Service*, 23–46, passim.

52 David A. Clary, *Timber and the Forest Service* (Lawrence: University Press of Kansas, 1986), 9.

53 Rodgers, *Fernow*, 237–38.

54 *USDA Annual Report for the Division of Forestry for 1886*, 165–66.

55 Steen, *Forest Service*, 37–46; and Kirkland, "American Forests," 94–97.

56 Congress, Senate. *Message from the President of the United States, Transmitting Report Relative to the Preservation of the Forests on the Public Domain*, 51st Cong., 1st sess., Ex. Doc. 36 (20 January 1890), 2.

57 Kirkland, "American Forests," 71–72; and Ron Arnold, "Congressman William Holman of Indiana: Unknown Founder of the National Forests," in Steen, *Origins*, 308. Who penned Section 24 remained the subject of much debate until Ron Arnold investigated it thoroughly. He concluded it was most likely Representative William Holman of Indiana. At the very least, Holman was responsible for the language of the bill. For the history of Section 24 and historiography relating to the bill, see Arnold, "Holman," 301–13.

58 Anthony Godfrey, *The Ever-Changing View: A History of the National Forests in California, 1891–1987* (R5-FR-001. San Francisco: USDA Forest Service, Pacific Southwest Region, 2005), 37–41. Requests for GLO investigations led to California's first forest reserves—San Gabriel Forest Reserve (1892), Sierra Forest Reserve (1893), San Bernardino Reserve (1893), and Trabuco Canyon Reserve (now part of the Cleveland National Forest) (1893).

59 Rodgers, *Fernow*, 220–21; and Steen, *Forest Service*, 32–33.

60 Congress, House. *Rational Forest Policy for the Forested Lands of the United States*, 54th Cong., 1st sess., HR 306 (17 March 1896). See also Steen, Forest Service, 29–30; and Cameron, Government Control, 205–209.

61 Sargent to Pinchot, 2 August 1890, Gifford Pinchot Papers, Manuscript Division, Library of Congress, Washington, DC; and Pinchot, *Breaking New Ground*, 32–33.

62 Sargent to Pinchot, 30 January 1896, Pinchot Papers.

63 Sutton, *Sargent*, 159; Harold T. Pinkett, *Gifford Pinchot: Private and Public Forester* (Urbana: University of Illinois Press, 1970), 14; and Martin Nelson McGeary, *Gifford Pinchot, Forester-Politician* (Princeton: Princeton University Press, 1960), 38. Pinkett and McGeary overlook Fernow's involvement in drafting the letter; McGeary interpreted the commission's work and its report as a battle between Sargent and Pinchot.

64 Pinchot, *Breaking New Ground*, 33; and Sutton, *Sargent*, 153.

65 Michael McCarthy, "The First Sagebrush Rebellion," in Steen, *Origins*, 190. Kirkland, "American Forests," covers in detail western outrage expressed in newspapers and local meetings, 229–38, and the various forest reserve bills brought before Congress in the aftermath of Cleveland's proclamation, 238–46.

66 Cox et al., *Well-Wooded Land*, 182.

67 Pinkett, *Public Forester*, 43–44.

68 Pinchot, *Breaking New Ground*, 130–31; Pinkett, *Public Forester*, 43; McGeary, *Forester-Politician*, 37–43; and Char Miller, "Before the Divide," in *Gifford Pinchot: The Evolution of an American Conservationist* (Milford, PA: Grey Towers Press, 1992) 15–16, 29–30.

69 Congress, Senate. Committee on Forest Reservations and the Protection of Game. *Forest Policy for the Forested Lands of the United States*. 55th Cong., 1st Sess., 25 May 1897, 10.

70 Senate. *Forest Policy*, 24–27. For more on the use of cavalry in the national parks, see H. Duane Hampton, *How the U.S. Cavalry Saved Our National Parks* (Bloomington: Indiana University Press, 1971).

71 Sutton, *Sargent*, 164.

72 Pettigrew noted that the forest reserves had been created in 1891 primarily for watershed protection, much to the opposition of some western irrigators. Their fight to prevent any lumbering in the reserves and grazing in the mountain ranges placed them at odds with foresters at the outset. However, they came to favor regulated lumbering because of the continued effectiveness of forested watersheds in moderating stream runoff. Irrigators then worked side by side with forestry supporters from that time forward, thereby bolstering Pinchot's work. Samuel P. Hays, *Conservation and the Gospel of Efficiency: The Progressive Conservation Movement, 1890–1920* (1959; reprint, Pittsburgh: University of Pittsburgh Press, 1999), 22–24.

73 30 *Statutes at Large*, 34–36 (4 June 1897).

74 30 *Statutes at Large*, 34–36 (4 June 1897); Steen, *Forest Service*, 34–36; and *Pinchot, Breaking New Ground*, 115–19. The "mark and remove" policy would haunt the Forest Service and force it to rethink its land management policies. The law, however, outlined federal forest management so clearly that it remained unchanged until passage of the Multiple Use–Sustained Yield Act of 1960, a law that supplemented—not replaced—the Organic Act.

75 Sutton, *Sargent*, 167.

76 Congress, Senate. *Surveys of Forest Reserves*, 55th Cong., 2nd sess., S. 189, 42; and Steen, *Forest Service*, 40, 51. For more on the debate about the Prussian and American roots of the Forest Service's administrative structure, see Ben W. Twight, "Bernhard Fernow and Prussian Forestry in America: Building Support for Federal Forestry," *Journal of Forestry* 88(2) (February 1990): 21–25; and Char Miller, "The Prussians Are Coming! The Prussians Are Coming! Bernhard Fernow

and the Roots of the USDA Forest Service," *Journal of Forestry* 89(3) (March 1991): 23–27, 42. Miller's article was a response to Twight's, and Twight responded to Miller in "'Fernow' Author Responds," *Journal of Forestry* 89(3) (March 1991): 25–27.

77 Congress, House. *Forestry Investigations and Work of the Department of Agriculture, 1877–1898*, by B.E. Fernow, 55th Cong., 3rd sess., HR 181, 5.

78 Fernow, *Forestry Investigations*, 3.

Chapter Two: Establishing the Forest Service

79 Miller, "All in the Family," 122.

80 Char Miller, *Gifford Pinchot and the Making of Modern Environmentalism* (Washington, DC: Island Press, 2001), 29–30.

81 Miller, *Gifford Pinchot*, 31.

82 Gifford's mother, Mary Eno Pinchot, was the eldest daughter of Amos R. Eno, a wealthy New York merchant who controlled more than $25 million worth of prime Manhattan real estate when he died in 1898. The Pinchot children inherited a good deal of this fortune.

83 Char Miller, "The Greening of Gifford Pinchot," in *Gifford Pinchot: The Evolution of an American Conservationist* (Milford, PA: Grey Towers Press, 1992), 52–53.

84 Edgar Brannon, Interview for *The Greatest Good: A Centennial Film*, transcript, U.S. Forest Service History Collection, Forest History Society, Durham, NC.

85 Char Miller, Interview for *The Greatest Good: A Centennial Film*, transcript, U.S. Forest Service History Collection, Forest History Society, Durham, NC; and Brannon, interview.

86 Pinkett, *Public Forester*, 16. One of Pinchot's professors at the Sheffield School of Science was William Brewer, author of the *Statistical Atlas of the Ninth Census* (1874), one of the first attempts at gathering data on forests in the United States. Brewer helped Franklin Hough with his forest investigation work and later worked with Pinchot on the National Forest Commission.

87 Brian Balogh, "Scientific Forestry and the Roots of the Modern American State: Gifford Pinchot's Path to Progressive Reform," *Environmental History* 7(2) (April 2002): 204–207.

88 Miller, interview.

89 Gifford Pinchot, Diaries, 5 January 1889, Pinchot Papers.

90 Pinchot, *Breaking New Ground*, 5; and Pinkett, *Public Forester*, 17.

91 Pinchot, *Breaking New Ground*, xv; and Pinchot, Diaries, 12 November 1889, Pinchot Papers.

92 Pinchot, *Breaking New Ground*, 7, 8–9.

93 Pinchot, Diaries, 12 April 1890, Pinchot Papers.

94 See Pinchot, Diaries, 1 January; 18 March; 24 March; 10 April; 28 and 29 April; 1 May; 2 June; 15 July; 22 August; and 2 September 1890, Pinchot Papers. He spent the first week of January with Brandis taking dictation "on the ways to get things done in America." Pinchot, Diaries, 1 January 1890, Pinchot Papers.

95 Pinchot to James Pinchot, 10 March 1890, Pinchot Papers.

96 Gifford Pinchot, "Forest-policy Abroad, Part 1," *Garden and Forest* 4 (14 January 1891): 21–22; Pinkett, *Public Forester*, 20; and Pinchot, *Breaking New Ground*, 34–35.

97 Pinchot, *Breaking New Ground*, 10–22. Pinchot, Diaries, 8 April 1890, Pinchot Papers, specifically discusses the many interests of the Forstmeister, and 30 April 1890, about Professor Landoldt planting a tree.

98 William B. Greeley, *Forests and Men* (Garden City, NY: Doubleday, 1951), 82.

99 Pinchot, *Breaking New Ground*, 10–22.

100 Pinchot to James Pinchot, 18 February 1890, Pinchot Papers.

101 Pinchot to James Pinchot, 31 August 1890, Pinchot Papers.

102 Pinchot to Mary Pinchot, 12 February 1890, Pinchot Papers.

103 Steen, *Forest Service*, 316; and Clary, *Timber and the Forest Service*, 46–47. Since the 1960s, critics of clearcutting have urged a return to selection cutting for both aesthetic and environmental reasons.

104 Pinchot, *Breaking New Ground*, 21, 30.

105 Pinchot, Diaries, 1890 December 22, Pinchot Papers.

106 Although the work carried out on the Adirondack League Club's holdings in the early 1890s was the first large-scale scientific forestry project in that region and possibly the United States, Alfred Dolge implemented forest management on his 29,000-acre holdings in the late 1870s. McMartin, *Great Forests of the Adirondacks*, 211. In 1914, Vanderbilt's widow sold most of the land to the federal government, which used it two years later to create Pisgah National Forest.

107 Pinchot to James Pinchot, 27 March 1893, and James Pinchot to Pinchot, 4 February 1891, 11 and 18 February 1892, 26 and 28 October 1892, Pinchot Papers. Even after taking over the Division of Forestry, Pinchot held true to his belief that "the forests must be regulated by principles different from those" regarded as essential in Europe. See Brandis to Pinchot, 7 October 1901, Pinchot Papers.

108 Pinchot Diaries, 7 November 1892; and Pinchot to Mary Pinchot, 10 November 1892; Charles McNamee to Pinchot, 4 August 1893, Pinchot Papers.

109 For more on the Biltmore School, see Carl Alwin Schenck, *The Biltmore Story: Recollections of the Beginning of Forestry in the United States,* edited by Ovid Butler (St. Paul: American Forest History Foundation and Minnesota Historical Society, 1955).

110 Daniel T. Rodgers, *Atlantic Crossings: Social Politics in a Progressive Era* (Cambridge, MA: Harvard University Press, 1998), 27. Rodgers noted that the following institutions were at one time or another housed in the United Charities Building: "Florence Kelley's National Consumers' League; the National Child Labor Committee; the National Housing Association; the New York School of Philanthropy; the Commission on the Church in Social Service of the Federal Council of Churches; the city's two largest private charitable relief agencies; the *Survey* magazine, clearinghouse for social policy news and debate of every sort; and a library used (as Franklin Roosevelt's secretary of labor remembered) by 'everybody who was in social work.'" Others like them were within a few blocks' radius.

111 Fernow to Pinchot, 1 July 1891, Pinchot Papers. William Seward Webb had approached Fernow about doing the survey work, but, Fernow told Pinchot, he was not yet ready to abandon his work in Washington. The following year, Fernow hoped to get the forestry contract for the Webb land but lost it to Pinchot. Dietrich Brandis to Pinchot, 26 May 1892, Pinchot Papers.

112 Pinkett, *Private and Public Forester*, 37.

113 Gifford Pinchot and Henry S. Graves, *The White Pine* (New York: The Century Co., 1896), p. vii, quoted in Pinkett, *Private and Public Forester*, 35. They conducted the study in central Pennsylvania and Franklin and Clinton counties, New York, with funding from D. Willis James, William E. Dodge, and James Pinchot.

114 Pinchot, *Breaking New Ground*, 134.

115 Pinchot, *Breaking New Ground*, 136.

116 Pinchot, *Conservation Diaries*, 93.

117 Pinchot, *Breaking New Ground*, 137.

118 Pinchot, *Breaking New Ground*, 135–36; Gifford Pinchot, *U.S. Department of Agriculture Report of the Chief of the Division of Forestry for 1898* (Washington, DC: Government Printing Office, 1898), 171; Pinchot, *Practical Assistance to Farmers, Lumbermen, and Others in Handling Forest Lands*, U.S. Division of Forestry Circular 21.

119 Quoted in Steen, *Forest Service*, 56–57.

120 Steen, *Forest Service*, 57–58.

121 Steen, *Forest Service*, 58.

122 Len Shoemaker, *Saga of a Forest Ranger: A Biography of William R. Kreutzer, Forest Ranger No.1, and a Historical*

Account of the U.S. Forest Service in Colorado (Boulder: University of Colorado Press, 1958), 16–17.

123 Pinchot, *Conservation Diaries*, 77, 98.

124 President Roosevelt's State of the Union Message, 3 December 1901, quoted in *Conservation in the United States: A Documentary History: Land and Water, 1900–1970* (New York: Chelsea House Publishers, 1971), 11.

125 Lewis L. Gould, *The Presidency of Theodore Roosevelt* (Lawrence: University Press of Kansas, 1991), 62.

126 McGeary, *Forester-Politician*, 61.

127 Quoted in Pinchot, *Breaking New Ground*, 261.

128 U.S. Department of Agriculture, *Report of the Forester for 1906* (Washington, DC: Government Printing Office, 1907), 14; and U.S. Department of Agriculture, *Report of the Forester for 1907* (Washington, DC: Government Printing Office, 1908), 13–14.

129 Steen, *Forest Service*, 53.

130 Pinchot, *Breaking New Ground*, 152.

131 Unpublished notes entitled "Milford," no date, Henry S. Graves Papers, Manuscript Division, Sterling Library, Yale University, New Haven, II-18, 213.

132 Nelson C. Brown, "Milford—The Cradle of American Forestry," *Yale Forest School News* 43(1): 8.

133 Steen, *Forest Service*, 83.

134 While dean at Yale, Graves taught William Greeley, the third chief; Robert Stuart, the fourth; and Gus Silcox, the fifth.

135 Pinchot, *Breaking New Ground*, 143 and 302. Price's suicide in 1914, brought about in part because he lost his job in 1910, weighed on Pinchot the rest of his life. Pinchot's contrition in his memoirs may be tinged by that.

136 Steen, *Forest Service*, 76.

137 Steen, *Forest Service*, 75–78.

138 Pinchot, *Breaking New Ground*, 182.

139 Steen, *Forest Service*, 81.

140 Quoted in Steen, *Forest Service*, 62.

141 Arthur C. Ringland, "Conserving Human and Natural Resources," 1970, Regional Oral History Office, University of California–Berkeley.

142 Circular 23, U.S. Department of Agriculture, Bureau of Forestry, 15 January 1902.

143 *The Use of the National Forest Reserves: Regulations and Instructions*, U.S. Department of Agriculture (Washington, DC: Government Printing Office, 1905), 17.

144 Quoted in Pinchot, *Breaking New Ground*, 266.

145 *The Use Book*, 24–25, 89; Robert J. Duhse, "The Saga of the Forest Rangers," *Elks Magazine* July–August 1986: 7, quoted in Terry L. West, *Centennial Mini-Histories of the Forest Service* (Washington, DC: USDA Forest Service, 1992), 32.

146 Frederick F. Moon, "Foresters of the Effete East" (speech, Alumni Reunion, Yale Forest School, New Haven, 21 December 1911), in *Alumni Reunion Yale Forest School* (New Haven: Yale University Press, 1912), 85–86; and Thorton T. Munger to D. S. Jeffers, 6 June 1910, "Experience Book" of the Yale Forestry Club Collection, Forest History Society Archives, Durham, NC.

147 Quoted in Steen, *Forest Service*, 62.

148 Pinchot, *Breaking New Ground*, 325–26. In 1905, Pinchot drafted the letter for Secretary Wilson's signature outlining the purpose of the Forest Service at the time of its establishment. In the letter, he says it is "in the long run," not the "longest time." Quoted in Pinchot, *Breaking New Ground*, 261.

149 William Cronon, Interview for *The Greatest Good: A Centennial Film*, transcript, U.S. Forest Service History Collection, Forest History Society, Durham, NC.

150 See Pinchot, *Breaking New Ground*, 319–26, for the story of its naming by Pinchot.

151 Pinchot, *Breaking New Ground*, 326. In the passage from pages 319–26, Pinchot discusses in detail his revelation that conservation of natural resources was the key to the future, and that because resources are the foundations of a nation's life, conservation could be the foundation of permanent peace among all nations: by conserving resources, a country conserves (or saves) its people, and this leads to peace.

152 Theodore Roosevelt, *Theodore Roosevelt: An Autobiography* (New York: Macmillan, 1913), 407.

153 Miller, "The Greening of Gifford Pinchot," 68–70.

154 Gould, *Theodore Roosevelt*, 206–207.

155 Brannon, interview.

156 Peter Pinchot, Interview for *The Greatest Good: A Centennial Film*, transcript, U.S. Forest Service History Collection, Forest History Society, Durham, NC.

Chapter Three: Powers, Policies, and Fires

157 A combination of factors, including overgrazing and disastrous winters that devastated the cattle industry in the mid-1880s, contributed to the rapid growth of the sheep industry and the decline of the cattle industry. By 1900, sheep outnumbered cattle in most western states.

158 Hays, *Gospel of Efficiency*, 57–58; Theodore Roosevelt, *Hunting Trips of a Ranchman: Sketches of Sport on the Northern Cattle Plains* (New York: G. P. Putnam's Sons, 1885), accessed at http://www.bartleby.com/52/4.html, 24 January 2005.

159 William D. Rowley, *U.S. Forest Service Grazing and Rangelands: A History* (College Station: Texas A and M University Press, 1985), 30–38.

160 Rowley, *Grazing and Rangelands*, 64–65; Hays, *Gospel of Efficiency*, 62–63.

161 Smaller stockmen also opposed the Roosevelt administration's plan for leasing large tracts of land to cattlemen for fear of being excluded from the national forests; farmers and other westerners opposed leasing because it would impede settlement. The Roosevelt administration eventually lost this battle, but not before the issue had aroused a great deal of anger in the West.

162 Steen, *Forest Service*, 88–89; Dana and Fairfax, *Forest and Range Policy*, 88–89, 376.

163 Steen, *Forest Service*, 89.

164 Steen, *Forest Service*, 79. The Forest Homestead Act reserved the agency's right to reject the claim if the land was deemed more valuable as timberland. Because settlers could acquire a streamwide strip up to one and one-half miles in length, which enabled one settler to control access to a valley, and because the most fertile farmland frequently lay along valley floors, where the most valuable timber usually grew, there was some opposition from within the agency to the act.

165 Hays, *Gospel of Efficiency*, 133–35.

166 See Michael McCarthy, "The First Sagebrush Rebellion: Forest Reserves and States Rights in Colorado and the West, 1891–1907," in *The Origins of the National Forests: A Centennial Symposium*, edited by Harold K. Steen (Durham, NC: Forest History Society, 1992), 180–97.

167 Miller, *Gifford Pinchot*, 164.

168 Gould, *Theodore Roosevelt*, 202–03.

169 Hays, *Gospel of Efficiency*, 137.

170 Pinchot, *Breaking New Ground*, 300.

171 Roosevelt, *Autobiography*, 404–405.

172 Quoted in Miller, *Gifford Pinchot*, 165–67.

173 See Pinchot, *Breaking New Ground*, 327–33; Roosevelt, *Autobiography*, 368.

174 Pinchot, *Breaking New Ground*, 331. The commission recommended the identification and initiation of the more pressing projects of conservation, and application of the principles of conservation before it was too late. It also called for a national waterways commission to draw up specific plans; such a commission was established by the River and Harbors Act of 1909.

175 McGeary, *Forester-Politician*, 97–98.

176 Pinkett, *Public Forester*, 116; Henry F. Pringle, *The Life and Times of William Howard Taft* (New York: Farrar & Rinehart, 1939), Vol. I, 474–77.

177 Pringle, *Taft*, Vol. II, 478; Pinkett, *Public Forester*, 116. Not only did Taft release Garfield but also he moved another Pinchot man, the assistant attorney general for the Department of the Interior, George Woodruff, to the post of federal judge in Hawaii. In addition, Department of Agriculture attorneys who had provided the basis for administrative waterpower policy found themselves transferred to the Solicitor's Office. Hays, *Gospel of Efficiency*, 150.

178 Pinchot, *Breaking New Ground*, 413–17; Elmo R. Richardson, *The Politics of Conservation: Crusades and Controversies, 1897–1913* (Berkeley: University of California Press, 1962), 65.

179 The two best studies of the controversy are Richardson's *Politics of Conservation* and James L. Penick, *Progressive Politics and Conservation: The Ballinger-Pinchot Affair* (Chicago: University of Chicago Press, 1968). For shorter summaries, see Paolo Coletta, *The Presidency of William Howard Taft* (Lawrence: University of Kansas Press, 1973), 91–98; and Hays, *Gospel of Efficiency*, 165–71.

180 Pinchot, *Breaking New Ground*, 447.

181 The Ballinger controversy opened the door to Theodore Roosevelt's return to politics. Pinchot and his brother Amos helped launch the Progressive "Bullmoose" Party and fund Roosevelt's run for the White House. Roosevelt and Taft split the Republican vote in the 1912 presidential election, leaving Democratic candidate Woodrow Wilson with an easy victory.

182 Quoted in Steen, *Forest Service*, 104.

183 Gifford Pinchot to James Pinchot, 16 May 1900, Pinchot Papers. Pinchot's relentless lobbying efforts on Capitol Hill led Senator Albert Beveridge to write to Gifford's mother, "Remember me, I beg, to His Majesty and my scape-grace brother Gifford—Lord! What trouble that wild rake gives us." Albert Beveridge to Mary Pinchot, 1 December 1903, Pinchot Papers.

184 Thomas Woolsey, Personal Narrative, 23 May 1912, Pinchot Papers.

185 Steen, *Forest Service*, 105.

186 Steen, *Forest Service*, 107–108.

187 The following section on research draws liberally from Harold K. Steen, *Forest Service Research: Finding Answers to Conservation's Questions* (Durham, NC: Forest History Society, 1998); Steen, *Forest Service*, 131–41; Gerald W. Williams, *The USDA Forest Service: The First Century* (Washington, DC: USDA Forest Service, 2000), 39–41; and John I. Zerbe and Phyllis A. D. Green, "Extending the Forest Resource: 90 Years of Progress at the Forest Products Laboratory," *Forest History Today* (Fall 1999): 9–14.

188 Steen, *Forest Service Research*, 8, 12–13; and Williams, *First Century*, 37.

189 Charles A. Nelson, "Born and Raised in Madison: The Forest Products Laboratory," *Forest History* 11 (July 1967): 6–14.

190 Pinchot, *Breaking New Ground*, 308.

191 Norman J. Schmaltz, "Raphael Zon: Forest Researcher," *Journal of Forest History* 24(1) (January 1980): 24–39; James G. Lewis, "Raphael Zon and Forestry's First School of Hard Knocks," *Journal of Forestry* 98(11) (November 2000): 13–17; and Char Miller, "Militant Forester: Raphael Zon," *Forest Magazine* Winter 2005: 22–23.

192 Steen, *Forest Service Research*, 7–9; and Kathleen L. Graham, *History of the Priest River Experiment Station* (RMRS-GTR-129, Fort Collins, CO: USDA Forest Service, Rocky Mountain Research Station, 2004), 14.

193 Steen, *Forest Service*, 137–38.

194 Steen, *Forest Service Research*, 12.

195 Steen, *Forest Service*, 137–41; Williams, *First Century*, 40.

196 Steen, *Forest Service Research*, 53.

197 Steen, *Forest Service Research*, 53–56.

198 Steen, *Forest Service Research*, 73, 77–83 passim.

199 Steen, *Forest Service*, 136; Pinchot, *Breaking New Ground*, 144.

200 Stephen J. Pyne, *Year of the Fires: The Story of the Great Fires of 1910* (New York: Viking, 2001), 78–81.

201 E. C. Pulaski, "Surrounded by Forest Fires: My Most Exciting Experience as a Forest Ranger," *American Forests* 23(356) (August 1923): 485–86. For more first-hand accounts of fighting the fire, see Hal Rothman, *"I'll Never Fight Fire with My Bare Hands Again": Recollections of the First Forest Rangers of the Inland Northwest* (Lawrence: University Press of Kansas, 1994).

202 Henry S. Graves, "Fundamentals of the Fire Problem," *American Forests* XVI(11) (November 1910): 629–30.

203 Pyne, *Year of the Fires*, 85.

204 F. A. Silcox, "How the Fires Were Fought," *American Forests* XVI(11) (November 1910), 631–39.

205 Pyne, *Year of the Fires*, 203–205.

206 Stephen J. Pyne, *Fire in America: A Cultural History of Wildland and Rural Fire* (Princeton: Princeton University Press, 1982), 79; Pyne, *Year of the Fires*, 101.

207 Steen, *Forest Service*, 129–30; Williams, *Americans and Their Forests*, 453.

208 W. B. Greeley, "Better Methods of Fire Control," *Proceedings of the Society of American Foresters* VI, No. 2 (1911): 165; Pyne, *Year of the Fires*, 262.

209 Williams, *First Century*, 39–41; Charles E. Hardy, *The Gisborne Era of Forest Fire Research* (Missoula, MT:

University of Montana Forest and Conservation Experiment Station, 1977), 12–13. For more on the Mann Gulch tragedy, see Norman Maclean, *Young Men and Fire* (Chicago, IL: University of Chicago Press, 1992).

210 Pyne, *Year of the Fires*, 264; Williams, *Americans and Their Forests*, 452–54.

211 Pyne, *Year of the Fires*, 85.

212 Pyne, *Year of the Fires*, 266–67.

213 Eugene Hester, "The Evolution of Park Service Fire Policy," *Renewable Resources Journal* 11(1) (Spring 1993): 14–15; and Stephen J. Pyne, "The Perils of Prescribed Fire: A Reconsideration," *Natural Resources Journal* 41 (Winter 2001): 1–8. For more on the history of change within Yellowstone following the fires of 1988, see *After the Fires: The Ecology of Change in Yellowstone National Park*, edited by Linda L. Wallace (New Haven: Yale University Press, 2004). For more on the Storm Canyon fires in 1994, see John N. Maclean, *Fire on the Mountain: The True Story of the South Canyon Fire* (New York: William Morrow, 1999).

214 Stephen Pyne, Interview for *The Greatest Good: A Centennial Film*, transcript, U.S. Forest Service History Collection, Forest History Society, Durham, NC.

215 Michael P. Dombeck, Christopher A. Wood, and Jack E. Williams, *From Conquest to Conservation: Our Public Lands Legacy* (Washington, DC: Island Press, 2003), 4–5, 128.

Chapter Four: Primacy of Timber

216 David A. Clary, *Timber and the Forest Service* (Lawrence: University Press of Kansas, 1986), 21.

217 Richmond L. Clow, "Timber Users, Timber Savers: The Homestake Mining Company and the First Regulated Timber Harvest," *Forest History Today* 1998, 6–11; and Greeley, *Forests and Men*, 59–60.

218 John Fedkiw, *Managing Multiple Uses on National Forests, 1905–1995: A 90-Year Learning Experience and It Isn't Finished Yet* (Washington, DC: USDA Forest Service, 1998), 14–15; and Williams, *First Century*, 54–55.

219 Quoted in Clary, *Timber and the Forest Service*, 34.

220 Robert E. Wolf, "National Forest Timber Sales and the Legacy of Gifford Pinchot: Managing a Forest and Making It Pay," in *American Forests: Nature, Culture, and Politics*, edited by Char Miller (Lawrence: University Press of Kansas, 1997), 92–93.

221 Quoted in Steen, *U.S. Forest Service*, 145.

222 George T. Morgan, Jr., *William B. Greeley: A Practical Forester* (St. Paul, MN: Forest History Society, 1961), 6–31 passim.

223 Steen, *Forest Service*, 108–109; and Morgan, *William B. Greeley*, 32–36.

224 Clary, *Timber and the Forest Service*, 29–30, 32.

225 Williams, *Americans and Their Forests*, 442–44.

226 Williams, *Americans and Their Forests*, 443; and Morgan, *William B. Greeley*, 36–37.

227 Steen, *Forest Service*, 111–12; Clary, *Timber and the Forest Service*, 24.

228 William B. Greeley, *Some Public and Economic Aspects of the Lumber Industry*, USDA Report 14, 1917, 13–100 passim; Greeley, *Forests and Men*, 118; and Steen, *Forest Service*, 113.

229 Clary, *Timber and the Forest Service*, 40.

230 Fedkiw, *Managing Multiple Uses*, 15.

231 Clary, *Timber and the Forest Service*, 48–50.

232 At the end of the war, Americans were operating eighty-one sawmills and producing two million feet of lumber a day. All forestry units merged into the Twentieth in October 1918. Accounts of the regiments' accomplishments, written by several commanding officers including Graves and Greeley, may be found in *American Forestry* 25(306) (June 1919).

233 Gerald W. Williams, "Women in the Forest Service: Early Historical Accounts" (unpublished manuscript, U.S. Forest Service History Collection, Forest History Society, Durham, NC), 3.

234 Quoted in Clary, *Timber and the Forest Service*, 70.

235 Clary, *Timber and the Forest Service*, 73; and Steen, *U.S. Forest Service, 187–88*.

236 Clary, *Timber and the Forest Service*, 73; and Miller, *Gifford Pinchot*, 282–85.

237 Clary, *Timber and the Forest Service*, 80–81.

238 News release, National Lumber Manufacturers Association, Box 123, Sustained Yield, National Lumber Manufacturers Association Collection, Forest History Society Archives, Durham, NC.

239 Steen, *Forest Service*, 223.

240 In the associative state, the "government compensates for its lack of administrative capacity by relying on the expertise of private entities—and particularly private corporations—to undertake difficult tasks. This process leads to the creation of an associative state in which the government facilitates private interactions yet lacks its own sources of power." The concept was embraced by Hoover and his successor, Franklin Roosevelt, in the first New Deal, but when the associative state failed to bring substantial relief to Americans in the mid-1930s, FDR went a step further in his second New Deal and established what became known as the welfare state. Quote from Edward D. Berkowitz, review of *From Warfare State to Welfare State: World War I, Compensatory State Building, and the Limits of the Modern Order*, by Mark Allen Eisner, *The American Historical Review* 106(3) (June 2001): 944. Ellis Hawley originated the term associative state in "Herbert Hoover, the Commerce Secretariat, and the Vision of an 'Associative State,'" *Journal of American History* 61 (June 1974): 116–40.

241 Quoted in Steen, *Forest Service*, 224.

242 Clary, *Timber and the Forest Service*, 35.

243 Steen, *Forest Service*, 199–202.

244 Steen, *Forest Service*, 202–03.

245 Steen, *Forest Service*, 218. For more on the shelterbelt, see R. Douglas Hurt, *The Dust Bowl: An Agricultural and Social History* (Chicago: Nelson-Hall, 1981) and Wilmon Henry Droze, *Trees, Prairies, and People: A History of Tree Planting in the Plains States* (Denton: Texas Woman's University, 1977).

246 Rowley, *Grazing and Rangelands*, 155–58.

247 Steen, *Forest Service*, 204–09.

248 Clary, *Timber and the Forest Service*, 96.

249 Steen, *U.S. Forest Service*, 227. In May 1935, the Supreme Court ruled in *Schechter Poultry Corporation v. United States* that NIRA was unconstitutional on the grounds that the law was an invalid delegation of legislative power to the president and exceeded the authority of the federal government to regulate intrastate commerce.

250 Clary, *Timber and the Forest Service*, 48–50.

251 Clary, *Timber and the Forest Service*, 100–101.

252 Quoted in Clary, *Timber and the Forest Service*, 101.

253 Nancy Langston, *Forest Dreams, Forest Nightmares: The Paradox of Old Growth in the Inland West* (Seattle: University of Washington Press, 1995), 191–92. See Langston, *Forest Dreams*, 157–200, for a history of the Forest Service cutting policies in Oregon's Blue Mountains region.

254 Clary, *Timber and the Forest Service*, 97–100.

255 Lester A. DeCoster, "Tree Farming Tenacity: Sixty Years Old and Still Going Strong," *Tree Farmer* 20(4) (July–August 2001): 6–15; *Forest Farmer*, "Forest Farm Association through the Years," *Forest Farmer* 50(6) (May 1991): 18–47; and Steen, *Forest Service*, 252–53.

256 George A. Garratt, "Wood in War and Peace," *Journal of Forestry* 42 (September 1944): 636–44.

257 J. J. Byrne, "Engineering on the Guayule Project," in *The History of Engineering in the Forest Service: A Compilation of History and Memoirs*, 1905–1989 (Washington, DC: USDA Forest Service, 1990), 447–57.

258 Williams, "Women in the Forest Service," 4; and Lee F. Pendergrass, "Dispelling Myths: Women's Contributions to the Forest Service in California," *Forest and Conservation History* 34(1) (January 1990): 23–24. On the Angeles National Forest, for example, forty-two of sixty-five AWS spotters were women; on the Los Padres, fifty-

one of ninety-one; the San Bernardino, forty-six of ninety-five; and on the Cleveland, twenty-nine of seventy-one. See William S. Brown, "Sky Watchers of the Hinterlands," *American Forests* 53 (December 1947): 554.

[259] See Bradley Biggs, *The Triple Nickles* (Hamden, CT: Shoe String Press, 1986), for more on the history of the battalion.

[260] Kathy Pitts, "Diary of a Conscientious Objector," *History Line* (Summer 1994): 8–12.

[261] Clary, *Timber and the Forest Service*, 129–30.

[262] Steen, *Forest Service*, 251–52; and Clary, *Timber and the Forest Service*, 129–30.

[263] Clary, *Timber and the Forest Service*, 131.

[264] Clary, *Timber and the Forest Service*, 113–14.

[265] Sally Fairfax, Interview for *The Greatest Good: A Centennial Film*, transcript, U.S. Forest Service History Collection, Forest History Society, Durham, NC.

Chapter Five: Recreation, Wilderness, and Wildlife

[266] Clary, *Timber and the Forest Service*, 150.

[267] L. C. Merriam, Jr., "Recreation," in *Encyclopedia of American Forest and Conservation History*, edited by Richard C. Davis (New York: Macmillan, 1983), 571–76.

[268] Robert W. Righter, *The Battle over Hetch Hetchy: America's Most Controversial Dam and the Birth of Modern Environmentalism* (New York: Oxford University Press, 2005), 6; Steen, Forest Service, 114–15.

[269] Steen, *Forest Service*, 114–15.

[270] Steen, *Forest Service*, 115–17.

[271] For more on Mather, see Robert Shankland, *Steve Mather of The National Parks* (3rd ed., New York: Knopf, 1970). Mather and Gifford Pinchot had much in common. Both were independently wealthy men who entered government service in hopes of making a difference for the country's natural resources. They both defended their agency's turf with a missionary zeal and a degree of inflexibility that at times made working with them difficult.

[272] Robert E. Wolf, *The Concept of Multiple Use: The Evolution of the Idea within the Forest Service and the Enactment of the Multiple-Use Sustained-Yield Act of 1960*, prepared for the Office of Technology Assessment of the U.S. Congress (Washington, DC: Contract Number N-3-2465.0), 17.

[273] Curt Meine, *Aldo Leopold: His Life and Work* (Madison: University of Wisconsin Press, 1988), 160.

[274] Frank A. Waugh, "Recreation Uses on the National Forests—A Report to Henry S. Graves," 1917, quoted in Steen, *Forest Service*, 120.

[275] Steen, *Forest Service*, 121–22.

[276] William B. Greeley, *Report of the Forester* (Washington, DC: Department of Agriculture, 1920), 4.

[277] Quoted in Harlan D. Unrau and G. Frank Williss, *Administrative History: Expansion of the National Park Service in the 1930s* (Denver: National Park Service, 1983), accessed at http://www.cr.nps.gov/history/ online_books/unrau-williss/adhi2a.htm.

[278] Quoted in Steen, *Forest Service*, 157.

[279] Steen, *Forest Service*, 152–62, passim.

[280] Unrau and Williss, *Administrative History*, wwwcr.nps.gov/history/online_books/unrau-williss/adhi4.htm.

[281] National Conference on Outdoor Recreation, *Recreation Resources of Federal Lands: Report of the Joint Committee on Recreational Survey of Federal Lands of the American Forestry Association and the National Parks Association* (Washington, DC: 1928), 54–55; and Aldo Leopold, "The Last Stand of the Wilderness," *American Forests and Forest Life* 31(382) (October 1925): 86–88, 91.

[282] *Recreation Resources of Federal Lands*, 75.

[283] Unrau and Williss, *Administrative History*, accessed at http://www.cr.nps.gov/history/online_books/unrau-williss/adhi2a.htm.

[284] Elmo R. Richardson, "Olympic National Park: Twenty Years of Controversy," *Forest History* 12 (April 1968): 6–15; Ben W. Twight, *Organizational Values and Political Power: The Forest Service versus the Olympic National Park* (University Park: Pennsylvania State University Press, 1983); and Unrau and Williss, *Administrative History*, accessed at www.cr.nps.gov/history/online_books/unrau-williss/adhi2c.htm.

[285] Quoted in Steen, *Forest Service*, 240. See Steen, *Forest Service*, 237–45, for more on the entire issue.

[286] Wolf, *Concept of Multiple Use*, 35.

[287] John C. Hendee, George H. Stankey, and Robert C. Lucas, *Wilderness Management*, 2nd ed., (Golden, CO: North American Press, 1990), 100.

[288] Steen, *Forest Service*, 154–55.

[289] The Forest Service approved Arthur Carhart's other recommendation, to leave the lake region of the Superior National Forest in northern Minnesota in primitive condition, in 1926. Accessible only by canoe, the Boundary Waters Canoe Area was dedicated in 1964. Paul S. Sutter, *Driven Wild: How the Fight against Automobiles Launched the Modern Wilderness Movement* (Seattle: University of Washington Press, 2002), 63–65. See also Donald Nicholas Baldwin, *The Quiet Revolution: Grass Roots of Today's Wilderness Preservation Movement* (Boulder, CO: Pruett, 1972) for more on Carhart's role in early wilderness policy.

290 For more on the Gila Wilderness Area, see Christopher J. Huggard, "America's First Wilderness Area: Aldo Leopold, the Forest Service, and the Gila of New Mexico, 1924–1980," in *Forests under Fire: A Century of Ecosystem Mismanagement in the Southwest*, edited by Christopher J. Huggard and Arthur R. Gómez (Tucson: University of Arizona Press, 2001).

291 Sutter, *Driven Wild*, 72.

292 James M. Glover, *A Wilderness Original: The Life of Bob Marshall* (Seattle: The Mountaineers, 1986), 79.

293 Marshall's time spent observing Eskimos and nonnatives in the tiny village of Wiseman led him to publish *Arctic Village* in 1933. The book in part celebrated a community that lived simply and harmoniously in a relatively unspoiled environment. His brother George later compiled notes from his trips to the Brooks Range into a well-received, posthumously published book, *Arctic Wilderness* (1956).

294 Steen, *Forest Service*, 209.

295 Robert Marshall, "The Forest for Recreation and a Program for Forest Recreation," from *A National Plan for American Forestry—A Report Prepared by the Forest Service*, U.S. Congress, Senate Document No. 12, Separate No. 6 (72nd Congress, 1933), 463, 471–79.

296 Steen, *Forest Service*, 211.

297 Paul W. Hirt, *A Conspiracy of Optimism: Management of the National Forests since World War Two* (Lincoln: University of Nebraska Press, 1994), 151.

298 Dombeck et al., *From Conquest to Conservation*, 28. Dombeck defines a recreation visitor-day as the equivalent of one person engaged in recreation for twelve hours, or two people for six hours, and so on.

299 Fedkiw, *Managing Multiple Uses*, 63.

300 Hirt, *Conspiracy of Optimism*, 158.

301 Hirt, *Conspiracy of Optimism*, 158–59.

302 Dennis Roth, "A History of Wildlife Management in the Forest Service" (unpublished manuscript, Forest History Society Archives, Durham, NC, 1988; revised and edited by Gerald W. Williams, 1989.)

303 Meine, *Aldo Leopold*, 147–48.

304 The Ecological Society of America, a nonpartisan, nonprofit organization of scientists, was founded in 1915.

305 Thomas R. Dunlap, *Saving America's Wildlife* (Princeton: Princeton University Press, 1988), 32–36, 42–43.

306 Dunlap, *Saving America's Wildlife*, 38–39.

307 Placement of the Bureau of Fisheries in the Commerce Department reflected its focus on commercial fishing. In 1939, the Bureau of Fisheries moved to the Department of Interior. A year later, it merged with Bureau of Biological Survey to become the U.S. Fish and Wildlife Service.

308 Roth, "History of Wildlife Management."

309 Quoted in Theodore Catton and Lisa Mighetto, *The Fish and Wildlife Job on the National Forests: A Century of Game and Fish Conservation, Habitat Protection, and Ecosystem Management* (Missoula, MT: Historical Research Associates for USDA Forest Service, 1999), 48. For two first-hand accounts of the deer drive, see Carmony, "Grand Canyon Deer Drive," 41–64.

310 The case *United States v. Chalk* stemmed from a disagreement on the Pisgah (North Carolina) National Game Preserve. To prevent overpopulation of white-tailed deer, the Forest Service authorized a hunt and wanted to collect license fees from North Carolina residents. The Forest Service also wanted to move some of the fawns to other national forests in the South. State officials resented what they viewed as an infringement of states' rights and eventually filed a lawsuit. See USDA Forest Service, "Historical Sketch of the Pisgah Federal Game Preserve, June 20, 1939," U.S. Forest Service History Collection, Forest History Society, Durham, NC.

311 For more on the history of wildlife management, see Catton, *Fish and Wildlife Job*; and *Wildlife and America: Contributions to an Understanding of American Wildlife and Its Conservation*, edited by Howard P. Brokaw (Washington, DC: Council on Environmental Quality, 1978).

312 Susan Flader, "Aldo Leopold's Legacy to Forestry," *Forest History Today* (1998): 3; and Meine, *Aldo Leopold*, 126–27. For an intellectual biography, see also Susan L. Flader, *Thinking Like a Mountain: Aldo Leopold and the Evolution of an Ecological Attitude toward Deer, Wolves, and Forests* (Madison: University of Wisconsin Press, 1974).

313 Meine, *Aldo Leopold*, 144–46.

314 Dunlap, *Saving America's Wildlife*, 90; and Meine, *Aldo Leopold*, 241–42.

315 For more on the impact of Elton on Leopold, see Meine, *Leopold*, 283–84. For more on Elton, see Peter Crowcroft, *Elton's Ecologists: A History of the Bureau of Animal Population* (Chicago: University of Chicago Press, 1991).

316 Dunlap, *Saving America's Wildlife*, 70–71, 90.

317 Aldo Leopold, *Game Management* (New York: Charles Scribner's Sons, 1933), 391.

318 Leopold, *Game Management*, 230. Each of the four groups had an interest in predatory animals.

319 Aldo Leopold, *A Sand County Almanac and Sketches Here and There* (New York: Oxford University Press, 1949), 130.

320 Quoted in Williams, *The First Century*, 63.

321 Congress passed three important fish and wildlife laws in 1934. The Fish and Wildlife Coordination Act tried to coordinate the development of a nationwide program of wildlife conservation and rehabilitation but did not contain adequate provisions for enforcement. The Migratory Bird Hunting Stamp Act, or Duck Stamp Act, authorized using receipts from hunting permits for purchasing and developing wetlands for the national refuge system. The Cooperative Wildlife Research Unit Program drew land-grant colleges and state agencies into fish and wildlife research. Leopold's friend and fellow wildlife advocate Jay "Ding" Darling was instrumental in the passage of the last two acts.

322 Quoted in Roth, "History of Wildlife Management."

323 Roth, "History of Wildlife Management."

324 Theodore Catton, "From Game Refuges to Ecosystem Assessments: Wildlife Management on the Western National Forests," *Journal of the West* 38 (October 1999): 40.

Chapter Six: Rise of the Environmental Movement

325 Hirt, *Conspiracy of Optimism*, xxii; and "Resources for Freedom," *Washington Office Information Digest* 65 (28 July 1952): 1.

326 Clary, *Timber and the Forest Service*, 162–63.

327 Critics have charged that the Knutson-Vandenberg Act of 1930, and others like it, encouraged cutting because it meant more operating funds for the Forest Service. Hirt, *Conspiracy of Optimism*, xxxiv.

328 Clary, *Timber and the Forest Service*, 157–58.

329 Dombeck et al., *From Conquest to Conservation*, 28. The issue of who controlled access to the national forest road system was not completely resolved until 1962, when Attorney General Robert Kennedy ruled that the secretary of Agriculture had the right to require that timber companies give the federal government reciprocal access to roads built across national forests.

330 Steen, *Forest Service*, 158–59.

331 Hirt, *Conspiracy of Optimism*, 225.

332 *Proceedings of the Fourth American Forestry Congress*, 160.

333 Steen, *Forest Service*, 305.

334 In the listed uses of the national forests, recreation supporters managed to have their interest listed first by placing the word "outdoor" in front of "recreation."

335 Steen, *Forest Service*, 298.

336 Keenan Montgomery and Clyde M. Walker, "The Clearcutting Controversy: A Forum on Clearcutting: Where Are We, and Where Are We Going?" *Journal of Forestry* 71(1) (January 1973): 11–12.

337 Mark W. T. Harvey, *A Symbol of Wilderness: Echo Park and the American Conservation Movement* (Albuquerque: University of New Mexico Press, 1994), 293–94.

338 Edward C. Crafts, "Foresters on Trial," *Journal of Forestry* 71(1) (January 1973): 15.

339 Hirt, *Conspiracy of Optimism*, 222–23, 225.

340 Ronald B. Hartzer and David A. Clary, *Half a Century in Forest Conservation: A Biography and Oral History of Edward P. Cliff* (Washington, DC: USDA Forest Service, 1981), 22.

341 Fedkiw, *Managing Multiple Uses*, 33–34.

342 Hirt, *Conspiracy of Optimism*, 234.

343 Lawrence Rakestraw and Mary Rakestraw, *History of the Willamette National Forest* (Eugene, OR: USDA, Willamette National Forest, 1991), 112.

344 Quoted in Hirt, *Conspiracy of Optimism*, 230.

345 Hirt, *Conspiracy of Optimism*, 231.

346 Fox, *American Conservation Movement*, 297; and Rachel Carson, *Silent Spring* (Boston: Houghton Mifflin, 1962), 116.

347 Clary, *Timber and the Forest Service*, 185.

348 Charles Connaughton, "Forestry's Toughest Problem," *Journal of Forestry* 64(7) (July 1966): 446–48; and "The Revolt against Clearcutting," *Journal of Forestry* 68(5) (May 1970): 264–65.

349 Thomas R. Dunlap, DDT: *Scientists, Citizens, and Public Policy* (Princeton: Princeton University Press, 1981), 97.

350 Hirt, *Conspiracy of Optimism*, 219.

351 Fedkiw, *Managing Multiple Uses*, 178–79.

352 Shelley Smith Mastran, and Nan Lowerre, *Mountaineers and Rangers: A History of Federal Forest Management in the Southern Appalachians, 1900–81* (Washington, DC: USDA Forest Service, 1983), 144.

353 Hartzer, *Half a Century in Forest Conservation*, 231.

354 Mastran, *Mountaineers and Rangers*, 144; and Fedkiw, *Managing Multiple Uses*, 33.

355 Hirt, *Conspiracy of Optimism*, 216.

356 Carolyn Merchant, *The Columbia Guide to American Environmental History* (New York: Columbia University Press, 2002), 181.

357 National Environmental Policy Act of 1970, Public Law 91–190, 83 stat. 852 (1970).

358 Merchant, *American Environmental History*, 181.

359 Robert W. Malmsheimer, Denise Keele, and Donald W. Floyd, "National Forest Litigation in the US Courts of Appeals," *Journal of Forestry* 102(2) (March 2004): 24.

360 Hirt, *Conspiracy of Optimism*, 232.

361 87 stat.; 16 USC, Section 2.

362 Jack Ward Thomas, *The Journals of a Forest Service Chief*, edited by Harold K. Steen (Durham, NC: Forest History Society and University of Washington Press, 2004), 24.

363 Max Peterson, Interview for *The Greatest Good: A Centennial Film*, transcript, U.S. Forest Service History Collection, Forest History Society, Durham, NC.

364 Harold K. Steen, *An Interview with F. Dale Robertson* (Durham, NC: Forest History Society, 1999), 81.

365 Stanley H. Anderson, "The Evolution of the Endangered Species Act," in *Private Property and the Endangered Species Act: Saving Habitats, Protecting Homes*, edited by Jason F. Shogren (Austin: University of Texas Press, 1998), 12–13.

366 Dombeck et al., *From Conquest to Conservation*, 106; and Gary C. Bryner, *U.S. Land and Natural Resources Policy: A Public Issues Handbook* (Westport, CT: Greenwood Press, 1998), 128. Bryner discusses how, in Senate testimony, Chief Jack Ward Thomas confirmed the agency's commitment to meeting quotas even though it opposed the timber salvage rider. See Chapter 9 for more on this.

367 Crafts, "Foresters on Trial," 14. Ed Crafts served as a deputy chief of the Forest Service in the 1950s, and then as director of the Bureau of Outdoor Recreation in the Department of Interior after 1962.

368 John Mumma, Interview for *The Greatest Good: A Centennial Film*, transcript, U.S. Forest Service History Collection, Forest History Society, Durham, NC.

369 Orville Daniels, Interview for *The Greatest Good: A Centennial Film*, transcript, U.S. Forest Service History Collection, Forest History Society, Durham, NC.

370 Orville Daniels, interview.

371 Robert E. Wolf, Interview for *The Greatest Good: A Centennial Film*, transcript, U.S. Forest Service History Collection, Forest History Society, Durham, NC. A complete oral history with Wolf is available through the Maureen and Mike Mansfield Library and the K. Ross Toole Archives, University of Montana, website. Wolf worked on the Anderson-Mansfield Reforestation Act of 1960, numerous Highway Act authorizations, the Youth Conservation Acts of 1959 and 1961, the Multiple Use–Sustained Yield Act of 1960, the 1962 Trade Act, the Wilderness Acts of 1964 and 1975, the Forest and Rangeland Renewable Resources Act of 1974, the National Forest Management Act of 1976, the 1978 Renewable Resources Extension Act, the 1978 Forest Research Act, and the 1978 Cooperative Forestry Assistance Act, among others.

372 Hartzer, *Half a Century in Forest Conservation*, 251.

373 Hirt, *Conspiracy of Optimism*, 249–51.

374 Gerald W. Williams, correspondence with author, 4 February 2005.

375 Quoted in Dombeck et.al., *From Conquest to Conservation*, 31.

376 Clary, *Timber and the Forest Service*, 186.

377 Dennis C. LeMaster, *Decade of Change: The Remaking of Forest Service Statutory Authority during the 1970s* (Westport, CT: Greenwood Press, 1984), 18.

378 Quoted in Arnold W. Bolle, "The Bitterroot Revisited: A University Re-View of the Forest Service," *Public Land Law Review* 10 (1989): 14. Bryce Pinchot was also quoted as saying that the clearcut "would have killed Father" or "would have killed the Old Man." Miller, *Gifford Pinchot*, 357.

379 James B. Craig, "Editorial—The Blow That Probably Wouldn't Have Killed Father," *American Forests* 79(5) (May 1973): 9.

380 Pinchot to Mary Pinchot, 12 February 1890, Pinchot Papers.

381 Al Wiener, "Gifford Pinchot Would Have Laughed," *American Forests* 79(11) (November 1973): 13.

382 Paul V. Ellefson, "Recommendations of Presidential Panel on Timber and the Environment," *Journal of Forestry* 71(10) (October 1973): 621–22; and Hirt, *Conspiracy of Optimism*, 249, 250.

383 Clary, *Timber and the Forest Service*, 187.

384 Hirt, *Conspiracy of Optimism*, 236–37.

385 Bolle, "The Bitterroot Revisited," 172.

386 Hartzer, *Half a Century in Forest Conservation*, 231.

387 Robert E. Wolf, interview.

388 Clary, *Timber and the Forest Service*, 192. The same year Congress passed NFMA, it passed a similar law for the Bureau of Land Management, the Federal Land Policy and Management Act (FLPMA).

389 Clary, *Timber and the Forest Service*, 192.

390 Dombeck et al., *From Conquest to Conservation*, 35–36.

391 Hirt, *Conspiracy of Optimism*, 263–64.

Chapter Seven: New Faces, Changing Values

392 "1906 Appointment," *Northern Region News* 16 (9 August 1977): 4.

393 Pendergrass, "Dispelling Myths," 18.

394 Julia T. Shinn, "The Ranger's Boss," *American Forests* 36(7) (July 1930): 459–60, 481.

395 Robert E. Wolf, interview.

396 Marian Leisz, Interview for *The Greatest Good: A Centennial Film*, transcript, U.S. Forest Service History Collection, Forest History Society, Durham, NC.

397 Ellie Towns, Interview for *The Greatest Good: A Centennial Film*, transcript, U.S. Forest Service History Collection, Forest History Society, Durham, NC.

398 Dombeck et.al., *From Conquest to Conservation*, 29–30.

399 Ben W. Twight, "Bernhard Fernow and Prussian Forestry in America," *Journal of Forestry* Vol. 88, No. 2 (February 1990): 21–25; Frank J. Harmon, "What Should Foresters Wear?: The Forest Service's Seventy-five Year Search for a Uniform," *Journal of Forest History* Vol. 24, No. 8 (October 1980): 188–199; and United States Department of Agriculture, Forest Service, Northern Region, "Major Kelley, Forest Service Pioneer, Retires After 38 Years," Biographical file, U.S. Forest Service History Collection, Forest History Society, Durham, NC.

400 Earl E. Clark, Interview with author, 31 May 2005. The 10th Mountain Division Foundation, a nonprofit organization that honors the memory of the group, is behind the construction of a dozen backcountry huts for use by skiers and hikers on Forest Service land in the Colorado Rocky Mountains. The hut system is a good example of public-private partnership between the Forest Service and private citizens and groups. For more on the 10th Mountain Division, see Flint Whitlock, *Soldiers on Skis: A Pictorial Memoir of the 10th Mountain Division* (Boulder, CO: Paladin Press, 1992). Several division members returned to Colorado after the war and launched the Colorado ski industry, which built over a dozen resorts on Forest Service land. For more on that, see Annie Gilbert Coleman, *Ski Style: Sport and Culture in the Rockies* (Lawrence: University Press of Kansas, 2004).

401 *Proceedings of the Fourth American Forestry Congress, October 29, 30, and 31* (Washington, DC: The American Forestry Association, 1953), 170.

402 Department of the Army, "An Army Study on Program Control in the U.S. Forest Service Department of Agriculture," (Washington, D.C.: Department of the Army), 1.

403 Edgar Brannon, interview.

404 Herbert Kaufman, *The Forest Ranger: A Study in Administrative Behavior* (Baltimore: The Johns Hopkins Press for Resources for the Future, Inc., 1960), 96, 161.

405 Kaufman, *The Forest Ranger*, x.

406 Doug Leisz, Interview for *The Greatest Good: A Centennial Film*, transcript, U.S. Forest Service History Collection, Forest History Society, Durham, NC.

407 R. W. Behan, "The Myth of the Omnipotent Forester," *Journal of Forestry* 64(6) (June 1966): 398.

408 Steen, *Forest Service*, 213–14.

409 L. G. Edmonds, *Lassie: The Wild Mountain Trail* (Racine, WI: Whitman Publishing, 1966), 52.

410 Jeff Lalande, "The 'Forest Ranger' in Popular Fiction, 1910–2000," *Forest History Today* Spring/Fall 2003: 18–19. My thanks to Dianne Timblin for pointing out Lalande's work on the Forest Service in fiction.

411 William Clifford Lawter, Jr., *Smokey Bear 20252: A Biography* (Alexandria, VA: Lindsay Smith Publishers, 1994), 294–96, 314–15.

412 Lida W. McBeath, "Eloise Gerry: A Woman of Forest Science," *Journal of Forest History* 22(3) (July 1978): 128–35.

413 Letters are quoted in Rosemary Holsinger, "A Novel Experiment: Hallie Comes to Eddy's Gulch," *Women in Forestry* 5 (Summer 1983): 21–25. The section on women lookouts and early rangers is based on James G. Lewis, "The Applicant Is No Gentleman: Women in the Forest Service," *Journal of Forestry*, forthcoming, July 2005.

414 John D. Guthrie, "Women as Forest Guards," *Journal of Forestry* 18(2) (February 1920): 151.

415 Williams, "Women in the Forest Service," 3.

416 Gladys Murray, "Impressions of a Lookout Job," *The Six-Twenty-Six* III(13) (November 1919): 7–8.

417 E. Cornell, "Extracts from Diary of Miss E. Cornell, Sanger Peak Lookout Station, Siskiyou National Forest, 1918," *The Six-Twenty-Six* III(12) (October 1919): 9–13.

418 Williams, "Women in the Forest Service," 3.

419 Williams, "Women in the Forest Service," 1.

420 Getrude Becker, Interview for *The Greatest Good: A Centennial Film*, transcript, U.S. Forest Service History Collection, Forest History Society, Durham, NC. Becker worked for nearly fifty years as a clerk and volunteer.

421 Doug Leisz, interview; and Getrude Becker, interview.

422 Quoted in Chuck James, "Women in the Forest Service: The Early Years," *Journal of Forestry* 89(3) (March 1991): 14–17; and Pendergrass, "Dispelling Myths," 24.

423 James, "Women in the Forest Service," 17.

424 "A Job with the Forest Service: Information about Permanent and Temporary Jobs with the U.S. Forest Service," (c. 1950), 3. In the U.S. Forest Service History Collection, Forest History Society, Durham, NC.

425 David E. Nye, *The History of the Youth Conservation Corps* (Washington, DC: USDA Forest Service, Human Resource Programs, 1980); Williams, First Century, 107, 116–17; and Marlette C. Lacey, "Job Corps Celebrates Silver—Goes for Gold," unpublished manuscript, "Job Corps 25th Anniversary" folder, U.S. Forest Service History Collection, Forest History Society, Durham, NC.

426 Frome, *The Forest Service*, 66–67.

427 Jacqueline S. Reinier, *An Interview with Geri Vanderveer Bergen* (Durham, NC: Forest History Society, Inc., 2001), 55–63. When Larson graduated from the University of California at Berkeley in 1962, she received the highest academic distinction ever bestowed upon a University of California forester to that time, receiving honorable mention for the University Medal, the highest student award given at commencement exercises, and was also elected to Phi Beta Kappa. *Forestry Education at the University of California: The First Fifty Years*, edited by Paul Casamajor (Berkeley: California Alumni Foresters, 1965), 206; and Michael Frome, *The Forest Service*, 2nd ed. (Boulder, CO: Westview Press, 1984), 68. Known as Geraldine Bergen Larson during most of her Forest Service career, after the death of her husband in 1987, she legally changed her name to Geri Vanderveer Bergen.

428 Reinier, 63–65; and Doug Leisz, interview. Larson's husband kept his business and they took turns commuting on weekends to see one another.

429 Michael Thoele, *Fire Line: Summer Battles of the West* (Golden, CO: Fulcrum Publishing, 1995), 139. Thoele has a chapter on what he calls "the sisterhood of wildland fire." The Forest Service has not kept good records on female firefighting crews, so it is not known with certainty when women were first hired as firefighters by the agency.

430 Deanne Shulman, "On Becoming a Smokejumper," *Smokejumper Magazine* January 2003, accessed at http://www.smokejumpers.com/smokejumper_magazine/item.php?articles_id=238&magazine_editions_id=14, 16 January 2005. Shulman and Owen never met in person. Owen offered Shulman support and guidance by mail and telephone during her first season until he was killed in a parachute accident in September 1981.

431 Frome, *Forest Service*, 66–67.

432 Godfrey, *Ever-Changing View*, 525.

433 Frome, *Forest Service*, 66.

434 *Affirmative Employment Plan Update*, 73; and Jennifer C. Thomas and Paul Mohai, "Racial, Gender, and Professional Diversification in the Forest Service from 1983 to 1992," *Policy Studies Journal* 23(2) (Summer 1995): 296–309.

435 USDA News Feature Service, "Against the Tide: How One Youth Won the Career of 'HIS' Choice," May 21, 1980, U.S. Forest Service History Collection, Forest History Society, Durham, NC.

436 "Against the Tide"; and "Chip Cartwright, 1994–1997," *Regional Foresters*, accessed at *Southwest Region: About Us* website http://www.fs.fed.us/r3/about/history/sw-rfs/cartwright.shtml, 29 July 2004.

437 Harold K. Steen, *An Interview with Michael P. Dombeck* (Durham, NC: Forest History Society, 2004): 7.

438 James J. Kennedy and Jack Ward Thomas, "Exit, Voice, and Loyalty of Wildlife Biologists in Public Natural Resource/Environmental Agencies," U.S. Forest Service Collection, Forest History Society, Durham, NC.

439 Quoted in Miller, "Militant Forester: Raphael Zon."

440 Thomas and Mohai, "Professional Diversification," 296–309.

441 Harold K. Steen, *An Interview with R. Max Peterson* (Durham, NC: Forest History Society, 1991 and 1992), 278.

442 Allen Spencer, "Brief History of the 1985 Black Class Complaint in the California Region of the Forest Service," accessed at http://www.mlode.com/~aspencer/class_1985_history_introduction.html, 10 January 2005.

443 In 1990, a group of male employees from Region 5 moved to intervene in the female employees' Title VII action. The district court denied their motion, holding that it was untimely. The court affirmed in an unpublished disposition (*Bernardi v. Yeutter*, 945 F.2d 408). Shortly thereafter, male employees brought a separate action against the Forest Service, challenging the terms of the consent decree. The district court dismissed that action, holding in part that the male employees could not bring an independent action challenging the terms of a consent decree (*Levitoff v. Espy*, WL 557674). The male employees petitioned for a writ of certiorari, which the U.S. Supreme Court denied. *Levitoff v. Glickman*, 117 S. Ct. 296 (1996). The cases are summarized in *Donnelly v. Glickman* (9716648), accessed at http://news.findlaw.com.

444 USDA Forest Service, 2003. "Region 5 Women's Settlement Agreement, Monitoring Council Report, First Report R[egion] 5 Women's Settlement Agreement (*Donnelly v. Veneman*)," U.S. Forest Service History Collection, Forest History Society, Durham, NC.

445 Frome, *Forest Service*, 65–66; Office of Personnel Management, *2000 Demographic Profile of the Federal Workforce*, Table 1, accessed at http://www.opm.gov/feddata/demogap/table1mw.pdf, 28 July 2004; and USDA Forest Service, *FY 2002 Affirmative Employment Plan (AEP) Update and FY 2001 Accomplishment Report for Women and Minorities* March 2002: 73.

446 "Field Notes," *Pacific/Southwest Log* March 1984: 7.

447 Steen, *Interview with F. Dale Robertson*, 76–77.

448 Arthur Bryant, Forest Service employee, Interview with author, 12 January 2005.

449 A. L. Richard, Forest Service employee, Interview with author, 10 January 2005; Gregory C. Smith, Forest

Service employee, Interview with author, 5 January 2005; Antoine (Tony) Dixon, Forest Service employee, Interview with author, 5 January 2005; and Harold K. Steen, *An Interview with Jack Ward Thomas*, (Durham, NC: Forest History Society, 2001), 81.

[450] Frome, *Forest Service*, 66; Mohai, "Professional Diversification," 296. By comparison, in 2000, Hispanics in the agency numbered 2,355, and American Indians, 1,693.

[451] Office of Personnel Management, *Federal Civilian Workforce Statistics Demographic Profile of the Federal Workforce as of September 2002*, 27, accessed at www.opm. gov/feddata/demograp/02demo.pdf, 1 February 2005.

[452] Tim Sullivan, "Transforming the Forest Service: Maverick Bureaucrat Wendy Herrett," *High Country News* 6 December 2004, accessed at http://www.hcn.org/servlets/hcn.Article?article_id=15160, 20 December 2004.

[453] "Field Notes," *Pacific/Southwest Log* March 1984: 8.

Chapter Eight: Traditional Forestry "Hits the Wall"

[454] Williams, *Americans and Their Forests*, 482; and Fedkiw, *Managing Multiple Uses*, 35–37.

[455] Donna M. Paananen, Richard F. Fowler, and Louis F. Wilson, "The Aerial War against Eastern Region Forest Insects, 1921–86," *Journal of Forest History* October 1987: 185.

[456] William G. Robbins, *Landscapes of Conflict: The Oregon Story, 1940–2000* (Seattle: University of Washington Press, 2004), 190–94; and Fedkiw, *Managing Multiple Uses*, 163.

[457] Robbins, *Landscapes of Conflict*, 194–95. TCDD, an impurity found in minute but detectable amounts in 2,4,5-T but not 2,4-D, was reclassified in 2001 by the Department of Health and Human Services as a "known human carcinogen." National Institute of Environmental Health Sciences press release, available at: http://www.niehs.nih. gov/oc/news/dioxadd.htm. My thanks to John Fiske and Ronald E. Stewart for helping to clarify the issue of pesticides and the Forest Service.

[458] Fedkiw, *Managing Multiple Uses*, 165.

[459] Robbins, *Landscapes of Conflict*, 199–205. Public authorities, chemical companies, and independent scientists immediately complained about the incomplete data and inaccuracies in the two Alsea studies. The EPA nevertheless issued an Emergency Suspension Notice of 2,4,5-T for forestry rights-of-way, and pastureland immediately upon the release of Alsea II in February 1979. Although contemporary studies revealed the Alsea studies were deeply flawed, the government decided reintroducing the herbicides was not worth the political price. Subsequent research by the National

Academy of Sciences supported the critics' claims that the Alsea miscarriage studies had "inadequate or insufficient evidence" to determine whether an association existed between exposure to herbicides and miscarriages. By then, however, chemical companies were no longer producing 2,4,5-T and it had disappeared from the U.S. market. See: The Committee to Review the Health Effects in Vietnam Veterans of Exposure to Herbicides, Division of Health Promotion and Disease Prevention, Institute of Medicine, *Veterans and Agent Orange: Health Effects of Herbicides Used in Vietnam* (Washington, D.C.: National Academy Press, 1994), 6, 42–43.

[460] Gary L. Larsen, "Herbicides, the Forest Service, and NEPA," *EPA Journal* 14(1) (January 1988): 38.

[461] Fedkiw, *Managing Multiple Uses*, 167; John Fiske, Interview with author, 1 February 2005.

[462] Fedkiw, *Managing Multiple Uses*, 167.

[463] Hartzer, *Half a Century in Forest Conservation*, 29. Chief Ed Cliff refused to speak about military or intelligence aspects of the Forest Service in Vietnam on record when he was interviewed about his career in 1980 because he believed the missions were still classified.

[464] Kenneth Conboy and James Morrison, *The CIA's Secret War in Tibet* (Lawrence: University of Kansas Press, 2002), 75. Conboy has written extensively about CIA operations in several Asian countries, including on Vietnam, Laos, and Indonesia, during the Cold War.

[465] Fred Brauer, Interview for *The Greatest Good: A Centennial Film*, transcript, U.S. Forest Service History Collection, Forest History Society, Durham, NC; and Conboy, *CIA's Secret War in Tibet*, 56–57.

[466] Fred Brauer, interview; and Conboy, *CIA's Secret War in Tibet*, 141.

[467] Conboy, *CIA's Secret War in Tibet*, 136–37.

[468] For more on the Forest Service and international forestry, see Terry West, "USDA Forest Service Involvement in Post World War II International Forestry" in *Changing Tropical Forests: Historical Perspectives on Today's Challenges in Central & South America*, edited by Harold K. Steen and Richard P. Tucker (Durham, NC: Forest History Society, 1992), 277–91; and Williams, *First Century*, 138–41.

[469] Paul Frederick Cecil, *Herbicidal Warfare: The Ranch Hand Project in Vietnam* (New York: Praeger Publishers, 1986), 29–35. The chemicals Agent Orange and Agent White received their names from the markings on their barrels. They had been developed to retard the growth of broad-leaved weeds and for the defoliation of crops such as cotton so that mechanical pickers could work more

efficiently, but the concentration levels used in Southeast Asia ensured maximum and prolonged effect on a broad range of jungle vegetation. Agent Blue was more effective on food crops, but only in dry weather because it was water soluable. Other agents included Green, Pink, and Purple, which were used between 1962 and 1964 before being replaced by Orange, Blue, and White. For more on this, see, Cecil, *Herbicidal Warfare*, 225–32; and William A. Buckingham, Jr., *Operation Ranch Hand: The Air Force and Herbicides in Southeast Asia*, 1961–1971 (Washington, DC: Office of Air Force History, U.S. Air Force), 195–202.

[470] Cecil, *Herbicidal Warfare*, 57–58; and Buckingham, *Operation Ranch Hand*, 109–112.

[471] Cecil, *Herbicidal Warfare*, 77–78.

[472] "U.S. Admits Move to Burn Forests," *New York Times*, July 22, 1972.

[473] Jay H. Cravens, *A Well Worn Path* (Huntington, WV: University Editions, 1994), 324; Buckingham, *Operation Ranch Hand*, 127; and Robert Reinhold, "A Modest Proposal—'Sherwood Forest,'" *New York Times*, July 23, 1972.

[474] Hartzer, *Half a Century in Forest Conservation*, 263.

[475] Jay H. Cravens, Interview with author, 23 June 2005.

[476] Barry R. Flamm and Jay H. Cravens, "Effects of War Damage on the Forest Resources of South Viet Nam," *Journal of Forestry* 69(11) (November 1971): 789.

[477] Cravens, *Well Worn Path*, 338.

[478] Cravens, *Well Worn Path*, 338–43.

[479] Cravens reproduced his detailed letters home to his family in *A Well Worn Path*, 112–454. In them, the reader can trace the change in Cravens's attitude from support of the war to disillusionment.

[480] J. Brooks Flippen, *Nixon and the Environment* (Albuquerque: University of New Mexico Press, 2000), 84–86.

[481] Flippen, *Nixon and the Environment*, 8–16.

[482] Quoted in LeMaster, *Decade of Change*, 18–19.

[483] Flippen, *Nixon and the Environment*, 196.

[484] Flippen, *Nixon and the Environment*, 197–98, 211.

[485] Mary J. Coulombe, "Exercising the Right to Object: A Brief History of the Forest Service Appeals Process," *Journal of Forestry* 102(2) (March 2004): 12.

[486] Gretchen M. R. Teich, Jacqueline Vaughn, and Hanna J. Cortner, "National Trends in the Use of Forest Service Administrative Appeals," *Journal of Forestry* 102(2) (March 2004): 16, 18.

[487] Sally K. Fairfax, "Riding into a Different Sunset: The Sagebrush Rebellion," *Journal of Forestry* 79(9) (August 1981): 516–20, 582; and Sally Fairfax, interview. In her interview Fairfax noted that the talk of transferring land to state and local control or selling it outright was "a bit of a hoax" because of its fiscal implications. A percentage of revenue generated from resource extraction on public land is returned to the locality by the federal government and the work is subsidized by federal money. If the land were turned over, which Fairfax contended no secretary of Interior would allow, either business expenses for extraction companies would go up or local taxpayers would have to make up the difference.

[488] R. McGregor Cawley, *Federal Land, Western Anger: The Sagebrush Rebellion and Environmental Politics* (Lawrence: University of Kansas Press, 1993), 3–4.

[489] Hirt, *Conspiracy of Optimism*, 272.

[490] Dombeck et al., *From Conquest to Conservation*, 37; Hirt, *Conspiracy of Optimism*, xxxviii–xxxix.

[491] Steen, *Interview with F. Dale Robertson*, 50–51. Max Peterson disagreed with Robertson about how his own retirement came about. He says he initiated it and that Robertson was one of the last to learn of it, whereas Robertson says that the decision came from within the Reagan administration and that he knew of Peterson's ouster before Peterson did. See Steen, *An Interview with R. Max Peterson*, 1992, U.S. Forest Service Collection, Forest History Society, Durham, NC: 177–78.

[492] Cheri Brooks, "Is FORPLAN Obsolete? An Interview with Norm Johnson," *Inner Voice* 7(5) (September–October 1995): 11–12.

[493] Richard M. Alston and David C. Iverson, "The Road from Timber RAM to FORPLAN: How Far Have We Traveled?" *Journal of Forestry* 85(6) (June 1987): 43–47.

[494] Brooks, "Is FORPLAN Obsolete?" 12.

[495] Hirt, *Conspiracy of Optimism*, xxxii–xxxiii and 274–75.

[496] Orville Daniels, interview.

[497] Hirt, *Conspiracy of Optimism*, xxxii–xxxiii and 274–75.

[498] Patricia Woods, Interview for *The Greatest Good: A Centennial Film*, transcript, U.S. Forest Service History Collection, Forest History Society, Durham, NC.

[499] Steen, *Interview with F. Dale Robertson*, 82.

[500] "On Speaking Out: Forest Supervisors' Feedback to the Chief," *Inner Voice* 2(1) (Winter 1990): 7, 11.

[501] "AFSEEE," *Inner Voice* 2(1) (Summer 1989): 1.

[502] Herbert E. McLean, "A Very Hot Potato," *American Forests* 96(3–4) (March–April 1990): 30–31, 65–67.

[503] The AFSEEE newsletter contains numerous articles discussing how employees were threatened with dismissal or transfer for speaking out. Jerry Franklin, who led the New Forestry efforts in the Pacific Northwest, said that a couple of forest supervisors and even a regional forester demanded his dismissal for promoting his alternative to existing management practices. Jerry

Franklin, Interview for *The Greatest Good: A Centennial Film*, transcript, U.S. Forest Service History Collection, Forest History Society, Durham, NC.

[504] Hirt, *Conspiracy of Optimism*, 285–86.

[505] *Interview with F. Dale Robertson*, 111–14.

[506] Hirt, *Conspiracy of Optimism*, 286–87; and "Unkindest Cut," *People Magazine*, July 6, 1992: 111.

[507] Stacie Oulton, "Missouri Foxtrotter: The Breed of Choice for More Than 20 Years," *Daily Sentinel* (Grand Junction, CO), August 21, 1994.

[508] Hirt, *Conspiracy of Optimism*, 286–87; and "Unkindest Cut," 111.

Chapter Nine: A New Land Ethic

[509] Aldo Leopold, "Ecological Conscience," quoted in Meine, *Leopold*, 499, 500.

[510] Aldo Leopold, "The Land Ethic," in *A Sand County Almanac and Sketches Here and There*, Special Commemorative Edition (New York: Oxford University Press, 1987), 204.

[511] The Forest Service's Pacific Northwest Research Station, Oregon State University, and the Willamette National Forest administer the Andrews Forest cooperatively. Funding for the research program comes from the National Science Foundation, Pacific Northwest Research Station, Oregon State University, and other sources. The forest is one of the twenty-four major ecosystem research sites in the United States funded through the National Science Foundation's Long-Term Ecological Research Program.

[512] Jerry Franklin, "Toward a New Forestry," *American Forests* 95 (November–December 1989): 37–44.

[513] Jerry Franklin, interview.

[514] F. Herbert Bormann, "Ecology: A Personal History," *Annual Review of Energy and the Environment* 21 (November 1996): 15.

[515] Richard C. Mallonée, "Mt. St. Helens Remembered...20 Years Later," *Logger and Lumberman Magazine* 49 (May 2000): 22–24. Rob Carson, in *Mount St. Helens: The Eruption and Recovery of a Volcano* (Seattle: Sasquatch Books, 1990), discusses efforts made by the U.S. Army Corps of Engineers, Weyerhaeuser and other timber companies, and the Forest Service to restore the mountain's forests and wildlife since the 1980 eruption.

[516] Paul E. Krauss to R. E. Worthington, 30 May 1980, File Folder "National Forests: Specific Forests—Gifford Pinchot NF, Mount St. Helens," in U.S. Forest Service History Collection, Forest History Society, Durham, NC.

[517] Charlie Crisafulli, Interview for *The Greatest Good: A Centennial Film*, transcript, U.S. Forest Service History Collection, Forest History Society, Durham, NC.

[518] Mike Crouse, "A Strong Case for Active Management of the Forests," *Loggers World* 30 (May 1994): 30, 36–37. To establish the monument on federal land only, the Forest Service acquired seventeen thousand acres from Weyerhaeuser in a land exchange.

[519] Jerry F. Franklin and James A. MacMahon, "Messages from a Mountain," *Science* 288(5469) (19 May 2000): 1183–84.

[520] Jerry Franklin, interview.

[521] Franklin, "Toward a New Forestry."

[522] Franklin, "Toward a New Forestry."

[523] William E. Shands, Anne Black, and James W. Giltmier, *From New Perspectives to Ecosystem Management: The Report of an Assessment of New Perspectives* (Milford, PA: Grey Towers Press for the Pinchot Institute for Conservation, 1993), 5.

[524] Susan Flader, "Missouri Pioneer in Sustainable Forestry," *Forest History Today* (forthcoming); and Donald Dale Jackson, "Leo's Legacy," *American Forests* 95 (September–October 1989): 80.

[525] Flader, "Missouri Pioneer."

[526] Steen, *Interview with F. Dale Robertson*, 82.

[527] Shands, *From New Perspectives to Ecosystem Management*, 1–2, 4–5. The report also looked at the introduction of New Perspectives on the Ouachita and Klamath national forests. In summer 1990, after Robertson had inspected clearcutting on the Ouachita, which covers 1.6 million acres in western Arkansas and eastern Oklahoma, Regional Forester John E. Alcock designated the Ouachita a lead New Perspectives forest, and Congress provided extra funds for projects and research. In contrast, the Klamath, located on 1.68 million acres in northern California and a small area in Oregon, had seen its timber program grind to a halt over the spotted owl controversy, resulting in severe budget and staff cuts. As a last resort to demonstrate that they could manage for resources other than timber, forest managers simply declared the Klamath a New Perspectives forest and began implementation without the kind of support provided to the Ouachita.

[528] Shands, *From New Perspectives to Ecosystem Management*, 7.

[529] Steen, *Interview with F. Dale Robertson*, 85–87.

[530] Franklin, "Toward a New Forestry," 38.

[531] Franklin, "Toward a New Forestry," 44.

[532] Fedkiw, *Managing Multiple Uses*, 184.

533 Hutch Brown and Gerald W. Williams, "Crossing the Divide: Forest Service Milestones, 1993–2000," unpublished manuscript courtesy of Gerald W. Williams, in author's possession.

534 For more on Forsman's research and the origins of the controversy, see Steven Lewis Yaffee, *The Wisdom of the Spotted Owl: Policy Lessons for a New Century* (Washington, DC: Island Press, 1994), 3–155.

535 Dombeck et al., *From Conquest to Conservation*, 38–39.

536 John Fedkiw, "The Forest Service's Pathway toward Ecosystem Management," *Journal of Forestry* 95(4) (April 1997): 33. Thomas's methodology, initially developed for integrating timber management with wildlife requirements in the Blue Mountains of eastern Oregon and southeastern Washington, was published in 1979 as USDA Agricultural Handbook No. 553, *Wildlife Habitats in Managed Forests: The Blue Mountains of Oregon and Washington*. Thomas was one of sixteen contributing authors and one of more than sixty scientists and natural resource professionals who contributed to the landmark book. Fedkiw, *Managing Multiple Uses*, 175–76.

537 For more on the conflict between loggers and environmentalists, see Terre Satterfield, *Anatomy of a Conflict: Identity, Knowledge, and Emotion in Old-Growth Forests* (Vancouver, B.C.: UBC Press, 2002).

538 Steen, *Interview with Jack Ward Thomas*, 76.

539 Robbins, *Landscapes of Conflict*, 210–11.

540 *A Conservation Strategy for the Northern Spotted Owl* (Portland, OR: Interagency Scientific Committee to Address the Conservation of the Northern Spotted Owl, 1990), 5.

541 *Conservation Strategy*, 2.

542 Dombeck et al., *From Conquest to Conservation*, 47.

543 Brown and Williams, "Crossing the Divide." About thirty percent of the area already had special congressional designations (such as wilderness areas and wild and scenic river areas). The remaining seventy percent was classified into late successional reserves (thirty percent), adaptive management areas (six percent), managed late successional areas (one percent), administratively withdrawn areas (six percent), riparian reserves (eleven percent), and matrix (sixteen percent). Matrix areas are federal lands outside of reserves, withdrawn areas, and managed late-successional areas.

544 Dombeck, *From Conquest to Conservation*, 48–9.

545 Jack Ward Thomas, *Jack Ward Thomas: The Journals of a Forest Service Chief*, edited by Harold K. Steen (Durham, NC: Forest History Society, in association with University of Washington Press, 2004), 205–08, 245–46, 310–12.

546 Brown and Williams, "Crossing the Divide."

547 Dombeck et al., *From Conquest to Conservation*, 109.

548 Steen, *Interview with F. Dale Robertson*, 72.

549 Steen, *Interview with Jack Ward Thomas*, 31.

550 Thomas, *Journals*, 63–65.

551 Dombeck et al., *From Conquest to Conservation*, 73.

552 Steen, *Interview with Jack Ward Thomas*, 65.

553 The law defined a salvage timber sale as "a timber sale for which an important reason for entry includes removal of disease or insect infected trees; dead, damaged or down trees; or trees affected by fire or imminently susceptible to fire or insect attack. Also included is the removal of associated trees or trees lacking characteristics of a healthy and viable ecosystem for ecosystem improvement or rehabilitation." Emergency Salvage Timber Sale Program (16 U.S.C.-1611 note, July 27, 1995.)

554 Steen, *Interview with Jack Ward Thomas*, 101.

555 Steen, *Forest Service*, xxvii–xxviii; and Steen, *Interview with Jack Ward Thomas*, 101.

556 Steen, *Interview with Jack Ward Thomas*, 64–66.

557 Steen, *Interview with Michael P. Dombeck*, 71–76.

558 Michael P. Dombeck, "A Gradual Unfolding of a National Purpose: A Natural Resource Agenda for the 21st Century" (speech, U.S. Forest Service, Washington, DC, March 1998), quoted in *Interview with Michael P. Dombeck*, 165.

559 Michael P. Dombeck, Interview with author, 6 January 2005.

560 Flippen, *Nixon and the Environment*, 197; and Dombeck et al., *From Conquest to Conservation*, 95–96.

561 Dombeck, *From Conquest to Conservation*, 95–96, 106.

562 Dombeck, *From Conquest to Conservation*, 103. As of 2005, there were 7,500 bridges to maintain in the national forests in addition to the 375,000 miles of roads. Twenty percent of the roads were open without restriction to all vehicle types (including passenger cars), fifty-eight percent were open to pickups and other high-clearance vehicles, and twenty-two percent were closed.

563 Steen, *Interview with Michael P. Dombeck*, 82.

564 Dombeck et al., *From Conquest to Conservation*, 106.

565 Steen, *Interview with Michael P. Dombeck*, 142–44.

566 Dombeck et al., *From Conquest to Conservation*, 115.

567 Dale N. Bosworth, Interview with author, 13 January 2005.

568 Bosworth's first job was on the St. Joe National Forest, now part of the Idaho Panhandle National Forests.

569 Dale N. Bosworth, "We Need a New National Debate" (speech, Izaak Walton League, Pierre, SD, 17 July 2003, accessed at http://www.fs.fed.us/news/2003/speeches/07/bosworth.shtml, 23 December 2004).

570 Dale Bosworth, interview.

571 Dale N. Bosworth, "Four Threats to the Nation's Forests and Grasslands" (speech, Idaho Environmental Forum, Boise, ID, 16 January 2004, accessed at www.fs.fed.us/news/2004/speeches/01/idaho-four-threats.shtml, 22 December 2004); and *Southern Forest Resource Assessment*, edited by David N. Wear and John G. Greis (Southern Research Station, 2002), 66.

572 Brown and Williams, "Crossing the Divide."

573 Bosworth, "Four Threats."

574 Juliet Eilperin, "New Rules Issued for National Forests: Some Environmental Protections Eased," *Washington Post*, 23 December 2004.

575 W. Brad Smith, Patrick D. Miles, John S. Vissage, and Scott A. Pugh, *Forest Resources of the United States, 2002* (St. Paul, MN: USDA Forest Service, North Central Research Station, 2004), 9.

576 Dale Bosworth, interview; and Dale Bosworth, "The Forest Service: A Story of Change" (speech, Centennial Forum, Boise, ID, 18 November 2004). My thanks to Hutch Brown of the Forest Service for providing this speech.

Chapter Ten: Reflections on the Greatest Good

577 Quoted in Lisa Cohn, "Searching For Harmony: The Forest Service's New Ecosystem Management," *Forest Perspectives* Autumn 1992 2(3): 7.

Index

A

Adirondack Forest Preserve, 11
Adirondack League Club, 17, 261
African American Strategy Group, 183
Agent Orange. *See* Herbicides
Alaska National Interest Lands Conservation Act, 243
Allen, E. T., 36, 42, 46, 61
American Antiquities Act of 1906, 238
American Association for the Advancement of Science (AAAS), 11, 18
American Forest Congress:
    of 1905: 38–39, 237;
    of 2005: 39, 243
American Forestry Association (*also* American Forestry Congress and American Forests), 11, 15–16, 76, 118–119, 237, 257–258, 266, 268, 270
American Forestry Congress. *See* American Forestry Association
American Forests. *See* American Forestry Association
American Forests, 76, 79, 123, 161, 257–258, 264, 266, 269, 272–273
American Tree Farm System, 103
Andrews Ecosystem Research Group, 208
Angeles National Forest (California), 110, 265
Apache National Forest (New Mexico), 131
Arbor Day, 9
Army Corps of Engineers, U.S., 71, 126, 247, 274
Arnold Arboretum, 14, 32
Association of Forest Service Employees for Environmental Ethics, 204

B

Ballinger, Richard, 64–67, 181, 263
Bartram, John, 258
Bartram, William, 258
Becker, Gertrude, 173, 270
Below-cost timber sales, 145
Bentham, Jeremy, xiii–xiv, 52
Bernardi, Gene, 177–178, 184, 242, 270
Beveridge, Albert, 261
Big Blowup. *See* Forest Service: fires of 1910
Biltmore Estate, 26, 31–32, 44
Biltmore Forest School, 32, 43, 237
Bitterroot National Forest (Montana): 37, 151;
    and terracing and clearcutting: 154–157, 159–160, 195, 202, 216, 234, 268
Bob Marshall Wilderness, 122

Bolle, Arnold: 156–160, 269;
    and panel and Bolle Report: 156
Boone and Crockett Club, 19, 38
Bosworth, Dale: 249, 275;
    as chief of Forest Service: xv, 85, 185, 227–231, 243, 249
Brandis, Dietrich, 28–32, 45, 261
Brannon, Edgar, 27, 55, 168, 260, 262, 270
Brauer, Fred, 191, 272
Brewer, William H., 15, 19, 258, 260
Bryant, Arthur, 183, 271
Bureau of Land Management (BLM): 126, 147, 184, 190, 216, 218–221, 224, 240, 248, 269;
    and fire: 83, 176;
    and logging: 157
Bureau of Outdoor Recreation, 240, 267
Bureau of Reclamation (Reclamation Service), 22, 35, 66, 126, 144
Bush, George H. W., 214, 218, 242
Bush, George W., 224, 228, 230–231
Butz, Earl, xi, 241–242

C

Camp Hale, 166, 191
Carhart, Arthur, 121–123, 125, 266
Carson, Rachel, 147–149, 187–188, 194, 240, 268
Cartwright, Charles "Chip," 179–180, 271
Central Intelligence Agency (CIA), 190–192
Chapman, Herman H., 78, 80
Chugach National Forest (Alaska), 65
Church, Frank, 157
Church Committee Guidelines, 157
Circular 21, 34, 36, 42, 48, 261
Civil Rights Act of 1964, 174, 177
Civil Service Reform Act of 1978, 184
Civilian Conservation Corps (CCC), 72, 80, 96, 98, 175, 212, 234, 239, 245
Civilian Public Service, 106
Clapp, Earle H.: 246;
    as chief of Research: 71–72, 97–98;
    as acting chief of Forest Service: 120, 240
Clark, William, 200
Clarke-McNary Act of 1924, 79–80, 93–94, 239, 245
Clary, David A., 91, 160, 255, 259, 261, 264–266, 268–269
Clean Air Act:
    of 1963: 149;
    of 1970: 72
Clean Water Act of 1972, 72, 85

Clearcutting:
  on private lands: 25, 95, 101, 103, 141;
  on national forests: x, 21, 95, 137–139, 151,
    160, 211;
  as silviculture practice: x, 10, 18, 30–31, 138,
    148, 156, 211, 214;
  Forest Service opposition to: 91–92, 95, 101,
    103, 215, 243;
  Forest Service support of: 91, 101, 139, 141,
    148–149, 157, 203, 212;
  public opposition to: x, 135, 140–141, 145,
    148–151, 154–161, 195, 203, 214, 219,
    242, 247, 261, 269, 274
Cleveland, Grover, 18–20, 233, 259
Cleveland National Forest (California), 259
Cliff, Edward P.: 247;
  as assistant chief: 141–142;
  as chief of Forest Service: 155, 159–160, 193,
    195, 240–241, 268, 272
Cline, McGarvey, 69
Clinton, William J., 219–224, 230, 243
Coconino National Forest (Arizona), 71
Coeur d'Alene National Forest (Idaho), 73
Colorado state legislature, 62
Collins, Sally, 184–185
Colville National Forest (Washington), 172
Combatologists, 180–181
Congressional Research Service, 147
Connaughton, Charles, 148–149, 268
Conservation movement, 9, 53, 62, 64, 68, 78,
    112, 132, 233, 255, 258, 260, 268
Conti, Samuel, 178
Coolidge, Calvin, 117, 239
Copeland Report. *See* National Plan for American
    Forestry
Council on Environmental Quality (CEQ), 152,
    241, 267
Coville, Frederick V., 57
Crafts, Edward C., 268–269
Cravens, Jay H., 193–194, 272–273
Cronon, William, 52, 262
Crowell, John, 200–201
Cutler, A. Rupert, 189

D
Daggett, Hallie Morse, 172–173
Daniels, Orville, 154, 202, 269, 273
Davis, A. P., 66
DeBonis, Jeff, 181, 204
DDT (dichloro-diphenyl-trichloroethane),
    147–149, 188, 190, 268
Department of Agriculture (USDA): 60, 183, 211,
    237, 239, 241, 260–263, 266;

beginnings of forestry in: xii, 9, 12, 16, 22, 25,
    32, 38, 40, 42, 47, 53, 57, 127, 233, 237,
    244;
  Forest Service as a division of: xii, 13, 34–36,
    38, 40, 42, 45, 87–88, 237, 270;
  and rivalry with Department of Interior: x, 22,
    36, 61, 100, 111, 114, 118–119, 240, 246
Department of Interior (USDI): ix, xiv, 33, 35–36,
    46, 53, 62, 68, 99–100, 127, 238–240, 244;
  and Division of Forestry in: x, xiii–xiv, 22–23,
    36, 49–50, 57, 237;
  and rivalry with Department of Agriculture:
    113, 119–120, 125, 240, 246
Desert Land Act of 1873, 10–11
Dinosaur National Monument, 144
District 1. *See* Region 1
Division of Forestry (USDA). *See* Forest Service
Division of Forestry (USDI). *See* Department
    of Interior
Division of Recreation and Lands, 124
Division of Timber Management, 158
Division of Wildlife Management, 134
Division R (USDI), 35–36, 50
Dolliver, Jonathan P., 65
Dombeck, Michael: 248–249, 255, 264, 267–271,
    273–275;
  as chief of Forest Service: 166, 224–228, 243;
  as acting director of Bureau of Land
    Management: 222, 224
Dowe, Helen, 173
Drey, Leo, 212–213
DuBois, Coert, 48
Dunnell, Mark, 12

E
Earth Day, 154, 194–195
Earth First!, 217
Earth Summit of 1992 (Rio Summit), 214, 244
Ecosystem Management:
  origins of: 207–209, 220, 233, 255, 267;
  Forest Service and: xiv, 202, 213–216,
    218–219, 222–224, 228, 235, 243, 248,
    274–275;
  on private lands: 212–213;
  and New Perspectives: 213–216, 221, 274
Egleston, Nathaniel H., 12–13, 17, 237, 258
Eisenhower, Dwight, 109, 126, 139, 145, 199
Elton, Charles, 132, 148, 208, 267
Emergency Salvage Timber Sale, 223, 243, 275
Endangered Species Acts (ESA): xi, 72, 196,
    217–218, 241–242, 247, 269;
  and impact on land management: 85, 147,
    152–153, 196, 217–218

Environmental movement, xi, 137–161, 194,
    199, 207, 257
Environmental Protection Agency (EPA), 188,
    194–195, 220–221, 242
Eveleth, Harriet, 163

F
Fall, Albert, x, 119
Fairfax, Sally K., 109, 255, 257, 263, 266, 273
Faulkner, Jeanine, 162, 177
Fernow, Bernard Eduard: 12, 14–16, 19, 21–22,
    29, 34, 45, 259–261, 270;
    as chief of Division of Forestry: 15–18, 20, 28,
        31, 69, 79, 100, 237, 259–260;
    as educator: 17–18, 22, 28, 42, 70, 140
Fire(s): 7, 22, 35–36, 51, 78, 87, 90–91, 93–94,
    99, 105–106, 115, 118, 145, 147, 163–164,
    168–169, 171–172, 176, 187, 191, 195, 199,
    228, 230, 234, 238, 243, 245;
    ecological role of: 7, 77–78, 208, 223, 230;
    suppression of: 76, 118, 124, 166, 177, 190,
        223;
    and logging: 109, 161, 208, 231;
    historical: 11;
    of 1910: 70, 73–75, 77–78, 89, 95, 163, 238,
        245, 265;
Fire protection: 23, 35, 87, 92, 95, 97–98, 103,
    105;
    Weeks Act on: 73, 79–80, 239;
    Clarke-McNary Act on: 79–80, 94, 239;
    and fire policy: 19, 21, 25, 234;
    See also 10 a.m. policy
Fish and Wildlife Service, U.S., 127, 147,
    152–153, 203, 216–219, 221, 224, 243, 267
555th Parachute Infantry Battalion. See Triple
    Nickles
Food, Agriculture, Conservation and Trade Act of
    1990, 242
Forest and Rangeland Renewable Resources
    Planning Act of 1974, 154
Forest Ecosystem Management Assessment Team
    (FEMAT), 219, 243
Forest Farmers Association, 103
Forest Homestead Act of 1906, 50, 59, 238, 263
Forest Management Act. See Organic Act
Forest Pest Control Act of 1947, 187, 240
Forest Products Laboratory, 68–69, 71–73, 104,
    132, 138, 171–172, 199, 238, 263–264
Forest Reserve Act of 1891, 15, 18–19, 59, 61–62,
    233, 258
Forest Reserve Manual, 36
Forest Service:
    in Agriculture: viii, ix, 9, 12, 13, 16, 22, 25,
        233, 237;

Hough as chief of: 13–14, 237;
budget: 13, 17, 22, 35, 45, 62, 68, 72, 73, 98,
    135, 139, 144, 154, 181, 195, 199, 200,
    202, 204, 222, 226, 227, 230, 248, 274;
Egleston as chief of: 13, 237, 258;
Fernow as chief of: 15–18, 20, 28, 31, 69, 79,
    100, 237, 259–260;
Pinchot as chief of: 34–43, 45–55, 57–66, 100,
    111, 173, 223, 233–234, 237–238;
and cooperative programs: 34, 35, 69, 72, 73,
    79, 92, 94, 99, 103, 107–109, 134, 239,
    240, 267, 269;
and national forest administration: 71, 121,
    258;
and range management: viii, 46–47, 50,
    57–59, 71, 88, 115, 119, 120, 130, 142,
    147, 181, 187, 196, 222, 234;
Graves as chief of: 66–68, 71, 76, 79, 85, 89,
    91–93, 112, 114–115, 117, 234, 238;
and research: 17, 68–73, 79–80, 104, 172, 181,
    192, 199, 208–212, 223, 225, 238, 239,
    240, 242, 245, 246;
and fire and fire policy: 51, 78–85, 98, 118,
    166, 187, 228, 230, 234;
and fires of 1910: 73–79, 87, 89, 95, 163, 238,
    245;
and Greeley as chief of: 79–80, 87, 90, 93–94,
    117, 120, 122–123, 140, 181, 234,
    238–239;
and timber regulation: viii, 50, 53, 60, 78,
    87–103, 107–109, 112, 127, 134, 135,
    137–147, 150–161, 166, 169, 187, 190,
    199, 201–202, 216, 217, 234, 238, 239,
    243, 274;
and World War I: 68, 87, 90, 91, 92, 166, 172;
and women employees: 92, 105–116,
    171–178, 181–183, 184–185, 235, 241,
    242, 265, 271;
Stuart as chief of: 95, 100, 239;
and multiple use: 91, 97, 101–102, 116, 120,
    125, 140, 141, 142–145, 157, 158, 160,
    169, 171, 201, 207, 214, 215, 234, 240;
and World War II: 84, 104–107, 125, 134, 137,
    172, 174, 180, 199, 234, 246;
Silcox as chief of: 80–81, 100–102, 120, 125,
    239–240;
herbicide and pesticide use by: 147, 148, 149,
    187–190, 191, 241;
and Cold War: 137, 139, 169, 190, 191, 202,
    234;
Watts as chief of: 103, 108–109, 120, 127, 138,
    240;
and National Park service rivalry: 87, 111–118,
    144, 171, 234, 245;

Forest Service, *continued*

and recreation: 87, 102, 111–114, 115–119, 120, 123–127, 140, 141, 142, 144, 145, 166, 167, 204, 225, 228, 229, 231, 234, 235, 240;

transfer threats against: viii, 118–120, 246;

Clapp as acting chief of: 120, 240;

and wilderness: viii, 53, 87, 111, 120–127, 127, 142, 143, 145–147, 226, 234, 239, 241, 242;

and wildlife management: 53, 72, 127–130, 134–135, 141, 142, 144, 152–154, 180, 204, 216, 274;

McArdle as chief of: 109, 127, 139, 142, 240;

and environmental movement: ix, 145–161, 178, 187–190, 195, 198–199, 202–205, 214, 216, 219, 223, 224, 226, 227, 231, 233, 235, 241;

Cliff as chief of: 155, 159–160, 193, 195, 240–241, 268, 272;

culture of: 50–51, 150–151, 154, 157, 159, 160, 163–171, 178, 180–185, 187, 202–205, 214, 217, 222, 224;

and military connection: 37, 92, 105, 106, 107, 164, 166–167, 190–194;

African Americans in: 163, 179–180, 183–184;

Hispanics in: 179–180, 183, 184, 272;

American Indians in: 179, 184, 272;

McGuire as chief of: 189, 196, 241–242;

and Vietnam War: 190–194, 272;

and forest planning: 85, 144, 152, 153, 160–161, 163, 195–199, 201–205, 212,213, 214, 220, 221, 226, 230, 231, 241;

Peterson as chief of: 153, 200–201, 225;

and ecosystem management: xii, 202, 207–224, 228–231, 233, 235, 243, 238, 273;

Thomas as chief of: 222–224, 226, 243;

and Emergency Salvage Timber Sale: 223–224, 243, 269, 275;

Dombeck as chief of: 166, 224–228, 243;

and Natural Resource Agenda: 224–228, 249;

Bosworth as chief of: xv, 85, 185, 227–231, 243, 249;

and Four Threats: 228–231, 243, 249

chronological history of: 237–243

Forest Service Manual, xiv, 168

Forestry: xiv–xv, 8, 9, 11–23, 25, 28–34, 38–39, 42–43, 45, 48–49, 53, 66, 68–70, 72, 76, 78–79, 88–89, 94, 97, 103, 109, 117, 125, 130, 132, 148, 151, 160, 178–181, 193–194, 199, 204, 211, 233, 235, 237, 239–240, 243–244, 255–260, 262, 265, 266, 268, 270–272;

and traditional utilitarian: 8;

and clearcutting (even-aged management): 154–156, 159, 211;

and selection cutting (uneven-aged management): 33–34, 211;

and ecosystem management: 130, 132, 134, 207, 209, 211–216

FORPLAN, 201–202, 273

Forsman, Eric D., 216–217, 274

Four Threats agenda, 228–231, 243, 249

Franklin, Jerry, 208, 211–212, 215, 273–274

Fulton, Charles W., 62

G

General Land Office (GLO), ix, 18, 23, 35–36, 41, 57, 64, 87, 99, 106, 115, 126, 237, 240

Geological Survey, U.S., 210

Gerry, Eloise, 171–172, 270

G. I. Bill, 167, 174

Gifford, Sanford, 26, 28

Gifford Pinchot National Forest (Washington), 209–210, 225, 242

Gila National Forest (New Mexico), 123, 239

Gila Wilderness Area, 122, 267

Gisborne, Harry, 79–80, 264

God Squad provision, 218, 242

Gore, Albert, 219, 222

Governors' Conference on Conservation, 64

Grant, Ulysses S., 12

Graves, Henry S.: 19, 22, 32–33, 67, 73, 78–79, 87, 90, 166, 181, 244, 261–262, 264–266; as associate chief: 34, 42, 49, 66–67; as chief of Forest Service: 66–68, 71, 76, 79, 85, 89, 91–93, 112, 114–115, 117, 234, 238; as dean of Yale Forest School: 43, 46, 89, 114, 260

Gray, Asa, 14

Grazing: 7, 9–11, 46–47, 51, 57, 89, 99, 112, 115, 117, 120–121, 123, 125, 130, 147–148, 187, 196, 235, 239–240, 250–251, 260, 262; Forest Service policies on: x, 53, 57–59, 71, 97, 99, 115, 119, 121, 142, 222, 234–235, 239; and Interior: 36, 46–47

Grazing Service, 100, 126, 199, 240

Greatest good, the concept, xiv–xv, 2, 42, 52–53, 58, 107–108, 112, 207, 215, 217, 228, 231, 233–235

Greeley, William B.: 43, 92–93, 120, 166, 245, 261, 264–265; as assistant forester: 89–90; as district forester: 46, 77, 79, 87, 89–90; as chief of Forest Service: 79–80, 87, 90, 93–94, 117, 120, 122–123, 140, 181, 234, 238–239;

and West Coast Lumbermen's Association: 94, 100–101
Greenpeace, 217
Grey Towers, 26–27, 43–45, 66, 241
Grimaud, Pierre, 58–59, 196

H
H. J. Andrews Experimental Forest, 208–209, 225
Harding, Warren G., x, 119
Harper, Vernon L., 72
Harrison, Benjamin, 18–19, 127
Healthy Forests Act of 2003, 85
Herbicide(s): 109, 272;
    and the Vietnam War: 190–194, 272;
    Forest Service use of: 148–149, 187–191;
    used in timber production: 109, 189–190;
    opposition to use of: 188–189;
    and Herbicide Wars: 189–190
Herrett, Wendy Milner, 177, 272
Herty, Charles H., 69
Hetch Hetchy controversy, 111–114, 266
Heyburn, Weldon B., 63
Hinckley (Minnesota), 73
Hirt, Paul, 125, 202, 255, 267–269, 273
Hitchcock, Ethan A., 35, 61–62
Hodel, Donald, 200
Holy Cross Forest Reserve (Colorado), 58
Homestake Mining Company, 87–88, 264
Homestead Act of 1862, 7, 10
Hoover, Herbert, 95, 239, 265
Hough, Franklin, 11–15, 28, 237, 260
Hubbard Brook Experimental Forest, 208–209
Hudson River School landscape painters, 7, 27–28
Humphrey, Hubert H., 145–147, 160, 241
Hunt, Richard Morris, 26
Hunt v. United States, 130

I
Ickes, Harold, x, 100, 119–120
Inland Waterways Commission, 63–64
Interagency Scientific Committee (ISC), 217, 242, 274
Interior Columbia Basin Ecosystem Management Project (ICBEMP), 220, 243
Izaak Walton League, 159, 275

J
Job Corps, 175, 241, 270
Johnson, Lyndon B., 175, 194, 202, 273
Jones, Alice Goen, 174
Journal of Forestry, 45, 70, 93, 260, 264–265, 268–270, 272–274

K
Kaufman, Herbert, 168, 184, 270
Kendrick, Mara, 177
Kennedy, Robert, 194, 268
King, Jr., Martin Luther, 194
Kirby Company, 34
Kirkland, II, Herbert Donald, 257, 259
Klamath National Forest (California), 172, 274
Knutson-Vandenberg Act of 1930, 139, 239, 268
Koch, Elers, 78, 80

L
Lake States Forest Experiment Station, 70
Lancaster, Bob, 176
Land-grant (1890s) schools, 183, 204, 240, 268
Landoldt, Elias, 29, 261
Lane, Franklin K., 114
Larson, Geraldine "Geri" Bergen, 175–176, 271
Lassie, 170–171, 270
Leavitt, Clyde, 46
Leisz, Doug, 164, 176, 270–271
Leisz, Marian, 164, 269
Leopold, Aldo: xiv, 181, 208, 253, 264–266;
    and "The Land Ethic": xiii, 207–208, 212, 215, 235;
    on recreation: 118, 122, 137;
    and wildlife management: 122–124, 130–134, 137;
    and Wilderness Society: 122–123, 239;
    writings of: xiii–xiv, 132–133, 146, 194, 207–208, 240, 266–267
Lewis and Clark National Forest (Montana), 150, 156
Light, Fred, 58–60, 196
Logan, Paul Howland, xvi, 180
Logging: 7, 10, 50, 68, 85, 90, 100, 169,174,186, 193–194, 208;
    on private lands: 88–90; 92, 103, 107–108, 240;
    on national forests: 87–89, 103, 107–108, 111, 113–114, 117, 119, 121, 124–125, 140, 143–144, 147–148, 150, 152–154, 156, 161, 198, 200–201, 211, 219–220, 222, 231, 234, 241
Loring, George B., 13
Los Padres National Forest (California), 106, 180, 265
Lowe, John, 233
Lumber codes. *See* Forest Service: timber policy
Lyng, Richard, 183

M
MacMahon, James A., 211, 274
Mark Twain Forest Watchers, 213

Mark Twain National Forest (Missouri), 212–213
Marsh, George Perkins: 9, 11–13, 15, 52, 233, 257–258;
    and *Man and Nature*: 8–9, 12, 237, 257–258
Marshall, Robert: 137, 146, 267;
    on recreation: 123–125;
    and Wilderness Society: 123, 239;
    as chief of Recreation and Lands: 124–125
Martynuik, Jennifer, 177
Mason, David T., 96, 100–101, 103–104, 107
Mather, Stephen, 115–117, 266
McArdle, Richard E.: 246;
    as chief of Forest Service: 109, 127, 139, 142, 240
McCarthy, M. H., 171–172
McCloud, Emma, 163
McGee, W J, 22, 52
McGuire, John R.: 247;
    as chief of Forest Service: 189, 196, 241–242
McIntire-Stennis Act of 1962, 72–73, 240
McKinley, William, 37, 48
McSweeney-McNary Act of 1928, 72, 80, 87, 239, 245
Meister, Ulrich, 29, 31
Messenger, Lori, xvi, 177
Metcalf, Lee, 156
Michaux, André, 258
Michaux, François André, 258
Miller, Char, xi, xiii, xv, 27, 255, 257, 259–260, 262–265, 269, 271
Mining, x, 10, 21, 37, 50, 53, 68, 77, 87–88, 112, 115, 121, 125, 147, 222, 237, 240
Mining Law of 1872, 21, 85, 237
Mission 66, 126, 171
Modoc National Forest (California), 106
Monongahela National Forest (West Virginia): 151, 158;
    and clearcutting: 151, 159–160, 195, 234, 241–242
Mount Hood National Forest (Oregon), 115, 198
Mount St. Helens eruption, 209–211, 242, 274
Muir, John: 19, 21, 57, 115, 123, 146, 208, 257–258;
    and Hetch Hetchy controversy: 111–112, 114
Multiple use: 52, 91, 97, 101–102, 111, 157–158, 169, 207, 212, 218, 253, 262, 264;
    as defense again National Park Service: 115–116, 120, 125;
    logging as essence of: 101–102, 142, 148;
    and marginality of nontimber resources: 143, 158;
    debates: 125;
    as policy to maintain management flexibility: 120, 143–145, 157, 160, 215;
    zoning concept: 141–142

Multiple Use–Sustained Yield Act of 1960: 127, 131, 142–144, 157, 187, 196, 234, 240, 247, 258, 260, 270;
    failure of: 143–144; 234
Mumma, John, 181, 205, 269
Murray, Gladys, 172, 270

N

National Conference on Outdoor Recreation, 117,120, 239, 266
National Conservation Association, 66
National Conservation Commission, 238
National Conservation Congress, 78
National Environmental Policy Act of 1969 (NEPA), xi, 85, 151–152, 157, 160, 176, 180, 190, 194, 196–197, 241, 247, 268, 272
National Fire Plan of 2000, 85, 230, 243
National Forest Commission, 18–22, 33, 260
National Forest Management Act of 1976 (NFMA), 72, 85, 146, 157, 160, 178, 180, 195, 197, 242, 247–248, 269
National Forest System, 4, 10, 17, 22, 40, 42, 50, 62–63, 79, 81, 97, 117, 185, 201, 204, 214–217, 226, 230, 248, 253, 258
National grasslands, 4, 41, 45, 71, 144, 247
National Industrial Recovery Act of 1933, 100, 239
National Lumber Manufacturers Association, 95, 101, 142, 237, 239, 265
National Park Association, 118
National Park Service: 106, 115–119, 144, 147, 203, 266;
    establishment of: 113, 115, 238;
    fire policy of: 81, 85, 242, 264;
    and rivalry with Forest Service: 87, 116–120, 171, 124, 126, 234, 245
National Plan for American Forestry (Copeland Report), 96–98, 123–124, 239, 246, 267
National Recovery Administration, 100
National Trails System Act of 1968, 147
Natural Resource Agenda, 224–225, 243, 249, 275
Natural Resources Defense Council, 157, 160
Nebraska National Forest (Nebraska), 100
Nelson, Gaylord, 194–195
New Deal: 72, 81, 96, 109, 126, 234, 265;
    forestry programs during: 80, 98–100, 119, 187
New Forestry, 209, 212, 214, 273–274
New Perspectives, 204–205, 213–214, 221, 242, 274
New York State College of Forestry, Cornell University, 22, 35, 237
Newell, Frederick, 22, 35, 52

Newlands, Francis G., 38
Nixon, Richard M.: 271, 273;
    and environmental movement: 151, 194–195;
    and President's Advisory Panel on Timber and the Environment: 158
Norcross, T. W., 140
Norris-Doxey Cooperative Farm Forestry Act of 1937, 99, 239
Northwest Forest Plan, 219–221, 243
Northern Spotted Owl, x, 153, 198, 216–218, 220–222, 230, 235, 243, 248, 274–275
Nuttall, Thomas, 258

O

Office of Management and Budget (OMB), 197
Old-growth forests, 159, 208, 213, 216, 221, 225, 274
Olmsted, Frederick E., 46
Olmsted, Frederick Law, 31
Ologists (or Combatologists), 180–181, 199
Olympic National Forest (Washington), 119
Olympic National Park, 266
Operation Outdoors, 126
Operation Pink Rose, 192–193
Operation Ranch Hand, 192, 272
Operation Sherwood Forest, 192–193, 272
Oregon: 62, 73, 92, 106, 17, 145, 184, 194, 206, 217, 221, 238–240, 248;
    multiple use in: 57, 102, 115, 220, 243, 247, 272, 274;
    Forest Service in: 105, 108, 145, 189–190, 198, 200, 208–109, 246, 265, 273
Organic Act of 1897, 15, 20, 22, 37, 47, 59, 71–72, 85, 87, 107, 112, 127, 142–143, 146, 157, 159–160, 178, 180, 195, 197, 233, 237, 240–242, 247–248, 258, 260, 269
Ouachita National Forest (Arkansas), 182
Owen, Allen "Mouse," 176, 271

P

Pearson, Gus, 71
Pesticide(s): 135, 137, 147–148, 189, 241;
    use on national forests: 148, 150, 166, 187–190, 247, 272
Peterson, R. Max: 247–248, 269, 271, 273;
    as chief of Forest Service: 153, 200–201, 225
Pettigrew, Richard, 20, 260
Pinchot, Constantine, 25
Pinchot, Cyrille Constantine Désiré, 25
Pinchot, Gifford: ix–xii, xiii–xv, 9, 18–23, 25, 27–28, 31–33, 67–71, 73, 78–79, 87, 89, 93–95, 103, 109, 111–112, 123, 139, 158, 160, 166, 199, 215, 235, 244–245, 253, 256–262, 264, 267;

education of: 25, 28–31;
    publications by: xv, 22, 32, 34, 36, 50, 253, 255, 257,–260;
    as chief of Forest Service: 34–43, 45–55, 57–66, 100, 111, 173, 223, 233–234, 237–238;
    and professional forestry: 25, 42–49;
    opposition to Forest Service transfer: 119–120;
    on conservation: 22, 38, 52–55, 61–66, 78, 112, 146, 208, 215, 223;
    on logging regulation: 59–61, 87–90, 93, 95–96, 109;
    and Ballinger: 64–67, 181;
    as icon in environmental movement: 146
Pinchot, Gifford Bryce, 157
Pinchot, James: 25–28, 43;
    and son Gifford: 25–28, 261, 263;
    and Yale Forest School: 43
Pinchot, Mary, 26, 260–261, 263, 269
Pinchot, Peter, 55, 262
Pisgah National Forest (North Carolina): 129, 261;
    and game preserve: 134
Pittman-Robertson Act of 1937, 134–135
Potter, Albert: 46–47, 57, 66, 71, 233;
    as chief of Grazing: 46–47, 58;
    as associate chief: 68
Prairie States Forestry Project (Shelterbelt Project), 99–100, 239
President's Materials Policy Commission, 137
Preston, J. F., 140
Price, Overton, 42, 89, 233
    as associate chief: 46–48, 66
Priest River Forest Experiment Station, 70, 79, 264
Public Lands Commission, 39, 64
Pulaski, Ed, 74–75
Pyne, Stephen, 85, 264

R

Ranger(s), xiv, 13, 24, 30, 36–37, 42–43, 45–46, 49, 50–52, 56, 58, 61, 74, 80, 84, 93–94, 106, 127, 130, 132, 137, 144, 149, 163–166, 168–174, 177, 179–180, 183–184, 196, 212, 226, 232, 240, 248, 260, 262, 264, 268–270
Reagan, Ronald: x, 199;
    administration of: 178, 200–201, 273;
    timber policy of: 201
Recreation: 72, 87, 97, 107, 111–113, 120–122, 124, 130, 131, 135, 137, 142–146, 148, 151, 178, 210, 218, 226, 228–229, 231, 234–235, 239–240, 253, 266, 268–269;
    budget: 122, 126, 143–144, 204;
    as economic asset: 112, 115–117;

Recreation, *continued*
emergence of in interwar period: 110,
115–116;
post-World War II boom: 135;
and logging in recreation areas: 107, 111, 117,
121, 123, 125–126, 140–144;
planning: 122–126, 225;
as promoted by Forest Service: 115–116, 126,
216;
staff growth: 126–127, 180;
visitor use: 126, 147–148
Region 2 (Rocky Mountain), 5, 41
Region 3 (Southwestern), 5, 41, 179
Region 4 (Intermountain), 5, 41, 228, 246, 249
Region 5 (Pacific Southwest), 5, 41, 174,
176–178, 180–182, 184, 204, 249, 269
Region 6 (Pacific Northwest), 5, 41, 177, 219
Region 7 (Northeast) 5, 41, 45, 241
Region 8 (Southern), 5, 41, 45, 184
Region 9 (Eastern), 5, 41, 45, 241
Region 10 (Alaska), 5, 41, 45
Rescission Act of 1995, 223, 243
Resources for Freedom, 137, 268
Rider, W. B., 171–172
Riley, Smith, 46
Ringland, Arthur, 49
Rio Grande National Forest (Colorado), 10, 71
Rise to the Future Fishery Program, 215–216
Roadless Area Review and Evaluation (RARE):
147, 195, 226, 241;
RARE II: 147, 226, 242
Roadless areas, 118, 123–124, 222, 225–228, 231,
243–244, 247, 249
Robertson, F. Dale: 221, 248, 269, 271, 273–275;
as chief of Forest Service: 153, 183, 204–205,
222, 224, 235, 242–243;
and ecosystem management: 204–205,
213–218, 221–222, 242–243
Roosevelt, Franklin D.: 99–100;
and New Deal: 98, 119, 263;
on Forest Service reorganization: 119–120;
conservation legislation of: 98, 100
Roosevelt, Theodore:
and Pinchot: ix, xiv, 22, 35, 38, 45, 54, 62, 64;
and conservation: 19, 38, 52, 54, 62–64, 112,
114, 146;
presidency of: 38–40, 42, 57, 60–64, 130, 199,
260–261
Rowley, Gordon, 177
Roth, Filibert, 35–36, 237, 246, 258

S
Sagebrush Rebellion:
of 1907: x, 59–60, 62, 257, 261
of 1980s: 199, 271

Samuel R. McKelvie National Forest (Nebraska),
100
San Bernardino National Forest (California), 118
Sand County Almanac, xiv, 207–208, 240, 267, 273
Sargent, Charles Sprague: 9, 13–15, 31–32, 237,
258–260;
and National Forest Commission: 19–22, 28
Schenck, Carl, 19, 32–33, 43, 46, 261
Schurz, Carl, 13–14, 19
Selection cutting: 95, 101, 158, 211, 261;
on national lands: 91, 95
Senior Executive Service, 205, 222
Shabecoff, Philip, 255
Shawnee National Forest (Illinois), 212–213
Shelterbelt Project. *See* Prairie States Forestry
Project
Shelton Cooperative Forest Agreement, 108
Shinn, Charles, 163–164
Shinn, Julia, 267
Shulman, Deanne, 176, 271
Sierra Club, xi, 112–113, 115, 142, 151, 237
Silcox, Ferdinand Augustus "Gus": 43, 48, 77–78,
245, 260;
as chief of Forest Service: 80–81, 100–102,
120, 125, 239–240;
and fire suppression: 79–81, 166, 264
Silent Spring, 147–149, 188, 194, 240, 268
Silviculture, 28, 31, 71, 79, 103, 138, 149, 243,
245, 251
Simpson Logging Company, 107–108
Ski industry, 125–126, 166–167, 270
Smith, Herbert "Dol," 39, 47–48, 66, 115, 140
Smith, Jr., Zane Grey, 190
Smokey Bear, 84, 169, 171, 192, 195, 243, 270
Society of American Foresters (SAF): 13, 45, 66,
78, 87, 93, 119, 125, 180, 237, 244, 264;
and forestry: 103, 233
Soil Conservation Service, 99, 189, 239
Soil erosion, 7, 9, 46, 57, 228, 234, 239
Southern Homestead Act of 1866, 7
Spotted owl. *See* Northern Spotted Owl
State and Private Forestry, Branch of, 71, 79, 127,
188, 246
Steen, Harold K., xv, 214, 218, 255, 257, 259,
261–275
Stuart, Robert: 43, 166, 245, 262;
as chief of Forest Service: 95, 100, 239
Sudworth, George, 71
Superior National Forest (Minnesota), 266
Sustained yield, 13, 31, 34, 86, 89, 96, 101–104,
127, 132, 139, 142, 161, 202, 265
Sustained-Yield Forest Management Act of 1944,
107, 240
Swift, Lloyd, 134–135

T

Taft, William Howard, 64–67, 114, 181, 238, 244, 263
Tahoe National Forest (California), 176
Taylor Grazing Act of 1934, 85, 99, 239
Taylor, Edward C., 99
10 a.m. policy, 81
10th Mountain Division, 166, 270
Term Permit Act of 1915, 115, 238
Thomas, Everett B., 47
Thomas, Jack Ward: 120, 218–219, 248, 255, 269, 271–272, 274–275;
    and northern spotted owl panels: 217–219;
    as chief of Forest Service: 222–224, 226, 243
Timber: x, 7–9, 11–14, 30, 33, 48–50, 52, 69, 73, 68–79, 85, 90, 92–92, 95, 97–98, 103, 111, 121–122, 135, 137–139, 148, 154, 157, 163, 166, 169, 178, 181, 187–188, 195, 197, 199–203, 205, 210, 212–214, 222–223, 226–227, 229, 234–235, 238–243, 255, 259, 261, 263, 268–269, 273–274;
    and logging regulation: 53, 87, 89–90, 92, 95–96, 100, 103, 109, 123, 137, 142, 147, 197, 230–231, 245, 247, 264;
    depredation of: 21;
    sale of: 7, 21, 32, 51, 59–60, 87–89, 91, 95, 103, 107, 139–140, 144–145, 161, 197–198, 200, 205, 213, 219, 223, 226, 246, 250–251;
    and Organic Act: 20, 47, 87, 127, 159;
    management of: 19, 87–92, 94, 102, 107–109, 112, 127, 134–135, 137–142, 150–151, 154, 156–158, 160, 180, 187, 190, 199, 204, 211–212, 214, 216–220, 223, 227–228, 234, 241, 243, 264, 274;
    and sustained yield: 20, 30–31, 89, 91, 96, 102–104, 107–109, 115, 139, 142, 161, 202, 234, 240;
    wartime demand for: 90, 95, 104, 109, 120, 125, 134, 246;
    and multiple use concept: 97, 102, 131, 135, 141–145, 154, 240;
    salvage: 114, 211, 223–224, 243, 246, 269, 275;
    and affect of logging on other multiple uses: 112, 114–115, 126, 142–145, 150–151, 157, 159–160, 201, 204, 208
Timber Conservation Board, 95, 100–101, 239
Timber Culture Act of 1873, 10–11
Timber industry, 13, 85, 93, 98, 127, 137, 142–143, 161, 188, 204, 208, 217–219, 222, 224, 231, 247
Timber Resources for America's Future, 187, 240
Timber salvage, 114, 161, 211, 223–224, 243, 246, 267

Tongass National Forest (Alaska), 94, 226, 242
Tongass Timber Reform Act of 1990, 243
Toumey, James, 43
Transfer Act of 1905, 42, 237
Tree Farm System, 103, 109
Triple Nickles (*also* 555th Parachute Infantry Battalion), 106–107, 266
Truman, Harry S, 120, 137
2, 4-D, 188
2, 4, 5-T, 188–189
Towns, Ellie, 165, 179, 270
Twenty-five percent fund, 59, 89, 197, 238

U

United Nations, 64, 199
U.S. Agency for International Development, 192–193
United States v. Chalk, 130, 266
United States v. Light, 58–60, 196
United States v. Grimaud, 58–59, 196
Use Book (also *Forest Service Manual*), 49–51, 168, 262

V

Vanderbilt, George, 26, 31–33, 261
Vietnam War: 1, 149–150, 188–189, 193–195, 272;
    Forest Service involved in: 190–194, 272
von Schrenk, Herman, 69

W

Wagon Wheel Gap Experiment Station, 71
Walcott, Charles, 34
Wallace, Henry, 100, 130
Warder, John A., 258
Water Quality Control Act of 1965, 149
Watershed protection, 8, 47, 57, 102, 155, 233, 258
Watt, James, 199–200
Watts, Lyle: 246;
    as chief of Forest Service: 103, 108–109, 120, 127, 138, 240
Waugh, Frank, 116, 264
Webb, William Seward, 261
Weeks Act of 1911, 41, 73, 79–80, 87
Wells, Philip P., 47, 58, 66
Wenatchee National Forest (Washington), 247
West Virginia, 82, 151, 157, 160–161, 189, 241–242, 248
Western Range, 100, 239
Weyerhaeuser Company, 188, 210–211, 274
White River National Forest (Colorado), 122, 166, 177, 191
Wiener, Al, 158, 269

Wild and Scenic Rivers Act of 1969, 147, 241
Wilderness, x, 52–54, 83, 87, 106, 110–112, 115,
    118, 120–131, 133, 135, 137, 142–143,
    145–147, 160, 178, 181, 184, 222, 225–226,
    230, 239, 241–241, 253, 258, 266–269, 275
Wilderness Act of 1964, xi, 145–147, 149, 151,
    157, 197, 226, 234, 241–242, 247, 269
Wilderness Society, 122–123, 142, 145, 151, 239
Wildlife, 7, 45, 53, 61, 72, 85, 110–111, 119,
    122–123, 125–135, 137, 141–145, 147–148,
    150, 152–153, 180, 189, 203–205, 216–218,
    222, 228–229, 240, 242, 247–248, 253, 255,
    266–268, 271, 274
Wildlife Society, 134
Wilds, Jetie, 183
Willamette National Forest (Oregon), 145, 204,
    209, 268, 273
Williams, Gerald W., 79, 84, 157, 169, 199, 255,
    263–265, 267, 269–270, 272, 274
Wilson, James, xiii, 22–23, 34, 37, 39, 42, 67, 262
Wilson, Woodrow, 112, 115, 263
Wolf, Bob, 147, 155–156, 160, 164
Women:
    in Forest Service: xv, 1, 163–164, 171–178,
        181–184, 234, 242, 265, 270–271;
    as district clerks: 92, 173–174;
    in World War II: 105–106, 174

Women in Natural Resources, 178
Woodruff, George W., 47, 58, 263
Woods, Patricia, 203, 272
Woodsy Owl, 195
Woolsey, Thomas, 67, 263
Workforce: 154, 204, 271–272;
    change in composition of: 177, 180–185
Works Progress Administration (*also* Works
Projects Administration), 72, 99, 245

Y

Yale Forest School, xiv, 14, 32, 43–45, 66, 77–78,
    89, 95, 114, 131, 233, 237, 244–245, 258, 262
Year of the Fires. *See* Forest Service: fires of 1910
Yellowstone Fires (1988), 83, 242, 264
Yellowstone National Park, 111, 242, 264
Yellowstone Timberland Reserve, 35, 127
Yosemite National Park, 21, 111–114
Young Adult Conservation Corps, 179, 242
Youth Conservation Corps, 146, 175, 241, 270

Z

Zon, Raphael: 69, 181, 264, 271;
    and research: 70–71;
    and Shelterbelt Project: 99